Knowledge Intensive Business Services

Knowledge Intensive Business Services

Organizational Forms and National Institutions

Edited by

Marcela Miozzo

Professor in Innovation Studies, Manchester Business School

Damian Grimshaw

Professor in Employment Studies, Manchester Business School

Edward Elgar
Cheltenham, UK • Northampton, MA, USA

Published by
Edward Elgar Publishing Limited
Glensanda House
Montpellier Parade
Cheltenham
Glos GL50 1UA
UK

Edward Elgar Publishing, Inc.
136 West Street
Suite 202
Northampton
Massachusetts 01060
USA

A catalogue record for this book
is available from the British Library

Library of Congress Cataloguing-in-Publication Data

Knowledge intensive business services : organizational forms and national institutions /
    Marcela Miozzo and Damian Grimshaw (editors)
        p. cm.
    Includes bibliographical references.
    1. Knowledge management. 2. Technological innovations. 3. Service industries. I. Miozzo, Marcela, 1963-   II. Grimshaw, Damian.

    HD30.2.K63655 2005
    658.4′062–dc22   2005049717

ISBN-13: 978 1 84542 236 3
ISBN-10: 1 84542 236 8

Printed and bound in Great Britain by MPG Books Ltd, Bodmin, Cornwall

Contents

Figures

Tables

Contributors

Steven Casper Keck Graduate Institute, California, USA

Damian Grimshaw Manchester Business School, University of Manchester, UK

Jeremy Howells Manchester Business School, University of Manchester, UK

Mark Lehrer Department of Management, University of Rhode Island, USA

Volker Mahnke Department of Informatics, Copenhagen Business School, Denmark

Marcela Miozzo Manchester Business School, University of Manchester, UK

Glenn Morgan Warwick Business School, University of Warwick, Coventry, UK

Bart Nooteboom Tilburg University, The Hague, Netherlands

Mikkel Lucas Overby Department of Informatics, Copenhagen Business School, Denmark

Serden Özcan Department of Informatics, Copenhagen Business School, Denmark

Sigrid Quack Social Science Research Centre, WZB, Berlin, Germany

Andrew Sturdy Tanaka Business School, Imperial College, London, UK

Sigurt Vitols Social Science Research Center Berlin (WZB), Germany

Acknowledgements

This book arises out of an international workshop held in the University of Manchester in November 2003. We are very grateful to fellow university colleagues Rod Coombs and Richard Whitley, who did not present papers but contributed to discussions on many of the issues presented in the book. We are also grateful to Leslie Willcocks who made a presentation and contributed to the debates. We would also like to thank Nirit Shimron and Mary O'Brien for their help in preparing the book.

Finally, we are grateful for permission to reproduce two chapters which appear in similar form in journals. Chapter 4 is published in similar form as M. Miozzo and D. Grimshaw (2005), 'Modularity and innovation in knowledge-intensive business services: IT outsourcing in Germany and the UK', *Research Policy*, **34** (9), 1419–39. We gratefully acknowledge permission from Elsevier. Chapter 6 is published in similar form as Grimshaw, D. and Miozzo, M. (2006), 'Institutional effects on the market for IT outsourcing: analysing clients, suppliers and staff transfer in Germany and the UK', *Organisation Studies*, forthcoming. We gratefully acknowledge permission from Sage Publications Ltd.

1. Knowledge intensive business services: understanding organizational forms and the role of country institutions

Damian Grimshaw and Marcela Miozzo

INTRODUCTION

Since the 1980s, OECD economies have become tertiary economies. Services account for more than 70 per cent of value-added and employment in OECD economies (OECD 2005), and, in most economies, services production growth far outstripped that of manufacturing during the 1990s (OECD 2003).[1] But it is the growth of a particular segment of the services sector – knowledge intensive business services – which has captured the attention of the media, the research community and policy-makers (EC 1998, 2003; Miles 2003; OECD 1999; Peneder et al. 2003).

These services involve the intensive use of high technologies, specialized skills and professional knowledge. Knowledge intensive business services (hereafter labelled as KIBS) are considered important because they represent an important source of job growth and value-added. Moreover, they act as potentially valuable intermediate inputs across a range of sectors of economic activity and thus directly shape the competitiveness and performance of manufacturing and service firms, as well as organizations administered by local and central government. Responding to claims that these services play a special role as drivers of economic growth, this book brings together scholars from a mix of disciplines to explore the nature and evolution of a range of KIBS – including computer services, management consultancy, R&D services and express delivery services.

As the chapters in this book document, KIBS share particular characteristics that to some extent distinguish them from the 'old' service economy. Like consumer services, they involve hard-to-measure, intangible activities (such as the provision of management consulting advice). But unlike other services activities, their delivery requires complex interactions between client organizations and service providers to facilitate transfer of

1

information and knowledge. Also, these services are often embedded in physical products and delivered through highly internationalized markets. Policy focus on the rise of these services to some extent reflects these features. For example, the European Commission has argued for the need both to improve the inputs to the production of business services (by expanding workforce skills, diffusing information and communication technologies and promoting R&D and innovation) and to strengthen the ability of client organizations to capture value-added from external services delivery by addressing problems of transparency, intangibility and quality (e.g. by promoting standards and appropriate documentation) (EC 2003).

This book seeks to contribute to such policy issues by focusing on two interrelated questions. First, what are the characteristics of inter-organizational collaboration (between KIBS firms and client organizations) and in what ways do they shape market growth and client performance? Second, what is the influence of a country's institutional arrangements on the evolution of markets for KIBS and the related organizational forms? The chapters highlight the diverse patterns of evolution of selected KIBS sectors, reflecting distinctive collaborative relations between KIBS firms and client organizations, variety in countries' institutional environments and the influence of multinational KIBS firms.

Because there are several competing definitions of KIBS, it is not possible to present a definitive picture of the pattern of KIBS in OECD economies. For Miles (2001a), KIBS encompass all those business services founded upon technical knowledge and/or professional knowledge. This broad definition captures both the social and institutional knowledge involved in many of the traditional professional services (such as management consultancy and legal services) and the emerging technological and technical knowledge involved in high-tech services (such as computer services and R&D services) (ibid.; see also, EC 1998, p. 6). However, many sectors are inevitably difficult to classify reliably since firms undertake a range of activities. For example, a logistics services firm may provide business with high-tech services, but in many cases its main activity may in fact be transport and ought not to be defined as KIBS (Miles 2001a, p. 5). Similarly, data on KIBS do not encompass the full range of knowledge intensive business *activities*, some of which are sold as business services while others are coordinated within the organization (Miles 2003). An exclusive policy focus on developing markets for KIBS is thus in danger of prioritizing external over internal KIBS provision in the absence of empirical evidence on the relative performance advantages of each mode of economic coordination.

It is not possible to provide a complete statistical picture of KIBS because of the difficulties of twinning an analytical definition of KIBS with sector definitions provided in standard datasets. Nevertheless, a focus on selected KIBS activities is informative. Figure 1.1 shows the patterns of relative growth in three KIBS activities (computing, R&D and other business services) compared to growth in services and manufacturing for eight selected OECD economies.[2]

During 1991–2001, average annual real production growth among KIBS far outpaced growth in total services (and total manufacturing) in all eight countries. Unfortunately no data were available for the USA and Japan. The pace of growth in KIBS tended to reflect the level of growth in total services, with France and Germany registering the lowest growth for both KIBS and total services (Figure 1.1). However, widely differing growth rates ought not to detract from the fact that for six of the countries shown KIBS constitute a relatively high share of total GDP – between 8 and 11 per cent. Indeed, slow growth in KIBS in France has not changed the fact that France ranked top in 1991 and 2001 in the size of KIBS as a share of GDP (see Appendix Table 1.2).

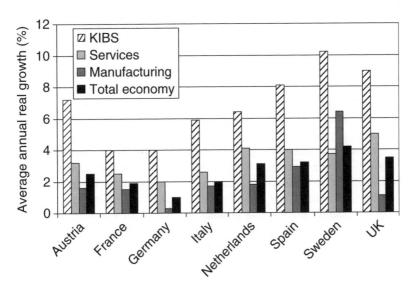

Note: KIBS is defined to include ISIC sectors 72 (computing and related), 73 (R&D) and 74 (other business activities). 1992–2001 data for Italy, 1995–1999 data for Spain, 1993–2001 data for Sweden and 1992–2000 data for the UK.

Source: see Appendix Table 1.1.

Figure 1.1   Growth of KIBS, total services, manufacturing and total economy in eight OECD countries, 1991–2001

If the 'Baumol disease' applied to these services, then their fast growth might be expected to dampen the growth rate of the total economy. Various studies suggest, however, that while Baumol (1967) concentrated on consumer services, a focus on KIBS offers a different conclusion since they are intermediary inputs for other industries and have the potential to raise productivity through knowledge spillovers, labour-saving, productivity and quality improvements. Therefore, as well as being an important potential source of growth in their own right, perhaps the more important feature of KIBS firms is that they are integrated into multiple stages of the value chain. According to the study by Peneder et al. (2003), KIBS substantially increased their share of intermediate inputs in OECD economies between the 1970s and 1990s. These shares increased from 17 per cent to 34 per cent in France, from 19 per cent to 31 per cent in the USA, from 5 per cent to 30 per cent in the UK, from 16 per cent to 26 per cent in Germany and from 8 per cent to 17 per cent in Japan (ibid.). Other studies have explored the impact of these inputs by analysing the performance of client industries. The fragmented evidence suggests a positive impact on value-added in client sectors reliant on business services (Antonelli's (1999) four-country European study), labour-saving and R&D spillovers to client sectors in the UK (Greenhalgh and Gregory 2000), and variable effects on client output and productivity contingent upon both country and sector (Di Cagno and Meliciani 2005; Drejer 2001; Tomlinson 2000).

Also, unlike other areas of private services (most notably consumer services), the very definition of KIBS implies that fast growth ought to be aligned with strong demand for highly educated workers and more intensive innovative activity – amounting to a strong potential for productivity growth. Evidence from the UK Community Innovation Survey-3 for 1998–2000 indicates a relatively intensive use of graduates among KIBS firms. Compared to firms in other services and manufacturing sectors, these firms were both more likely to hire graduates and to employ a higher proportion of graduates to non-graduates. Of all sectors, the only sectors where 90 per cent or more of firms employed graduates and where the median share of graduates was 50 per cent or more were computer services, R&D and technical testing, and architecture and engineering services (Tether and Swann 2003). Regarding innovation, again, while services account for a small share of total business expenditure on R&D (the EU average in 2001 was 13 per cent), levels of activity in KIBS firms can be expected to be higher. Data from the European Community Innovation Survey show that while reported innovative activity and R&D were low in services compared to manufacturing, figures for computer services and technical services were among the highest, suggesting stronger commonalities with high-tech manufacturing sectors than with other services:

72 per cent of innovating computer service firms and 58 per cent of technical services conducted R&D (the comparative figures are 47 per cent and 69 per cent, respectively, for all services and manufacturing) (Miles 2001b, p. 25). Also, technology-related KIBS appear to be better linked to innovation systems than services in general (Miles 1999).

Many factors have combined to generate and shape growth in knowledge intensive business services. In part, their growth is a result of wide-ranging, and interdependent, structural changes that have been identified as underpinning the shift, or 'gear change' as David and Foray (2003) call it, to a knowledge-based economy: a growing internationalization; a rise in education; and a technological system based on information and communication technologies (Petit 2002). The present phase of internationalization is distinctive in the range and scope of international transactions and increasingly, these cover services, as well as manufacturing (Miozzo and Miles 2002). While many accounts emphasize the cost competition from less developed economies as a threat to developed economies' efforts to sustain key sectors, other studies argue for the need to build comparative advantage around the development of KIBS and their linkages with sectors manufacturing quality products (Peneder et al. 2003). At the same time, however, the high incidence of multinational corporations among KIBS sectors makes the strategies of firms more relevant and may make it difficult for national or regional policy-makers to coordinate the development of economy-wide production systems.

The long-term rise in education, with more years of compulsory schooling and higher participation in tertiary education across all advanced economies, is viewed in many accounts as providing KIBS firms with the necessary supply of skilled human capital to meet their demands. Indeed, the broader category of 'business services' generated more jobs for highly educated workers than any other sector for the EU during 1995 to 2000 (EC 2003, Figure E).[3] Productivity gains from use of sophisticated technologies, high knowledge intensity, complex interaction with clients and reliance on tacit knowledge are said to be contingent upon firms having access to a pool of highly educated workers (EC 2003). Following Reich's (1990) recommendations, investment in tertiary education, continuous learning, and the updating and mobility of skills have been of special importance to economies' efforts to grow the pool of 'symbolic analysts'. Without such efforts, there have been fears, within countries and regions, that increasingly footloose internationalized KIBS firms will seek alternative locations that offer a more conducive human capital infrastructure.

These structural changes have interacted with the rapid diffusion of information and communication technologies, as characterized by the development of the PC and broadband Internet technologies. The digital revolution has transformed the way firms do research and develop new products and processes, as well as changed forms of delivery and consumption. New technologies have also increased the complexity of products and services in all sectors of economic activity, increasing the knowledge intensity of a range of manufactured products and services (Coombs 1999). In particular, growth in KIBS has benefited from the rapid expansion of the codified knowledge base, enabling the transmission of information across long distances (thus aligning with the internationalization of KIBS firms) and within complex networks at high speed and low cost (Cohendet and Meyer-Krahmer 2001). Indeed, information technology has allowed for the increased transportability of service activities, particularly those which have been most constrained by the geographical or time proximity of production and consumption (Miozzo and Soete 2001). However, the intangibility of services impinges a particular character on to this increased transportability. It means that much of the new trade enabled by new information technologies will be in the form of intra-firm rather than arm's-length transactions. Indeed, clients will tend to purchase services from local branches that belong to a worldwide network and to exchange services with their headquarters by means of computer-to-computer communications technology (Sapir 1987).

A further significant driver of the growing demand for KIBS, and one that is a focus of several chapters in this book, concerns changes in firm strategy and in the institutional environment. Outsourcing of services activities previously carried out in-house has increased the demand for specialized services provision. And, with the growth of the market for contracted-out services, external providers have become increasingly specialized (following Stigler 1951). Table 1.1 illustrates the linkages between the different management functions that have been traditionally administered in-house and the range of markets for business services that have grown as a result of outsourcing these functions. For completeness, the list includes both knowledge intensive and operational business services. There are obvious difficulties in putting a figure on the size of the overall outsourcing market; one study estimates a value for 2003 of €26.4 billion in Europe and €38 billion in the USA (www.tpi.net).

Outsourcing is a practice found across all sectors of economic activity and reflects changes in the institutional environment, as well as new modes of business strategy that favour concentration on so-called core capabilities. In the public sector, privatization and 'marketization' – in response to political imperatives (and to some extent changing macroeconomic constraints on governments) – have been especially important in driving growth in outsourcing. Evidence in Europe of a growing trend in the use of so-called 'mega-contracts' during 2000–04 was in large part fuelled by the expanding

Table 1.1   Linkages between internal management functions and markets for business services

Internal management function	Related market for business services
Administration	Management consultancy Legal services Auditing and accountancy
Human resources	Temporary work agencies Personnel recruitment Professional training
Finance	Banking Insurance Renting and leasing
Production and technical function	Engineering and technical services Tests and quality control R&D services Industrial design Maintenance and repair of equipment
Information systems	Software and IT services Telecommunications
Marketing and sales	Advertising Distributive trade Public relations Fairs and exhibitions After-sales services
Transport and logistics	Logistics Transport services Express courier services
Facility management	Security services Cleaning services Catering Environmental services/waste disposal Energy and water services Real estate (warehouses)

Source: adapted from EC (2003, Annex 1, Box 1).

public sector market for business services. It is notable that despite the absence of empirical evidence on value-added from externalization, public sector organizations still face strong pressures from national and pannational governmental bodies. For example, communications within the European Commission call for the 'modernization of public administration' through encouraging public–private partnerships, with the strong and unsupported claim that this generates 'gains in efficiency and also lower costs to the user as a result of competitive pressure' (EC 2003, p. 25).

Unlike the private sector, in policy debates regarding outsourcing of public services it is difficult to separate out the political goals (lessening the size of government and shifting responsibility for the delivery of public services) from the strategic business goals (exploiting the potential value added from specialization). This is not to say that strategy among private sector firms is characterized by an apolitical scientific approach, only that it avoids the strongly political issue of public versus private ownership, central to defining distributional issues in mixed capitalist economies. Other political issues are present in private sector firms' outsourcing strategies. These reflect financial pressures to downsize and cut costs, as well as opportunities to externalize groups of highly unionized workers, or workers whose strong professional identity (and strong wage demands) may be at odds with the firm's human resources strategy.

Given that outsourcing is a major driver of KIBS, what remains unclear in the literature is the degree to which growth in KIBS from outsourcing largely represents merely a change in the division of labour, with a substitution of externally provided business services for internal coordination of production. Such substitution has clear statistical consequences – one of which is a deflation of the estimated size of the manufacturing sector compared to the business services sector. In the UK, for example, a surprise collapse in the manufacturing sector in 2001 (a fall of 5.4 per cent) led to governmental calls to review traditional classifications of sectoral contribution to GDP and a response by the Trade and Industry Secretary that manufacturing activity would have been one-third higher had outsourced activities been retained in-house (*Financial Times*, 24 January 02).

Substitution also changes the political economy of production. This is especially notable where externalization of business services extends the presence of multinational KIBS firms in the domestic economy, or where private sector KIBS providers take on the role of delivering public services that were previously managed and delivered within a public sector organization. But if the expansion of KIBS (as measured in the form of distinctive industry classifications) is largely a substitution effect, then policy ought to concentrate on developing and supporting in-house as well as externalized business services.

Several studies have sought to assess the extent to which a shift to external services provision generates gains in value-added arising from possible increasing returns from the specialization of production – following Smith's argument about the division of labour (Arora and Gambardella 1994). But such studies are limited by the extent to which they can compare the relative value-added of marketized business services as opposed to internally coordinated business services with a method that controls for other extraneous factors. Other studies argue that whether or not specialization through outsourcing brings value-added, many areas of KIBS are sold to client organizations as innovative, complementary activities, not as a substitution for activities developed in-house (Gallouj and Weinstein 1997). Organizations may decide that the knowledge required for fast-changing technologies, or for acquiring and implementing innovations in products and processes, is inappropriate for in-house development, either because it is too costly, too far removed from core capabilities, or below a minimum efficient scale (Coombs 1999). Outsourcing may thus be a means of acquiring new knowledge that complements, rather than simply substitutes for, the existing portfolio of business activities.

Thus, there are powerful forces that are driving the development of knowledge intensive business services. This book seeks to fill some of the gaps in the literature through empirical and theoretical analysis of KIBS. The chapters respond to several claims – that the expansion of KIBS is considered fundamental to sustainable patterns of economic development, that their evolution is strongly associated with new 'network' organizational forms, and that KIBS organizations provide a potentially important source of innovation in the economy as well as generating spillovers for other organizations. What is distinctive about the contributions in this book is their attention to inter-organizational relations between clients and KIBS firms, strategies of multinational KIBS firms and the effect of changes in countries' institutional arrangements. The arguments and empirical evidence deepen our understanding of the complex evolution of KIBS in response to the changes in structural conditions and the interrelationship with patterns of innovation, inter-organizational relations and organizational performance.

OVERVIEW OF THE BOOK

The chapters in this book were selected from papers presented at an international workshop hosted at the ESRC Centre for Research on Innovation and Competitiveness (CRIC) at the University of Manchester in November 2003. The idea for the workshop was motivated by a need to fill gaps in our knowledge about the factors shaping the growth of knowledge intensive business services and the consequences for client firm innovation and performance.[4] Data or services sector activities have highlighted the rapid growth in many areas, as we have shown above. But they cannot reveal the complex conditions underpinning market expansion, nor can they answer questions about the extent to which there are national varieties or a universal pattern of the nature and evolution of knowledge intensive business services. Also, while there have been considerable advances in understanding the reasons for differential performance among firms, we identified a need to address the implications for theories of the firm posed by the apparent changes in boundaries, control and coordination of business services provision. As such, the book is strongly oriented towards both policy and theoretical questions.

The chapters are organized into two parts. The first part, 'Knowledge intensive business services and organizational forms', includes four chapters. In Chapter 2, Bart Nooteboom provides an integrated summary of theoretical principles concerning competence and governance and identifies several propositions for inter-organizational forms of collaboration. Chapter 3, by Jeremy Howells, argues that outsourcing R&D cannot be theorized as simply a make or buy decision, since it often involves long-term relationships and interactive, multiple links within networks of firms. In Chapter 4, Marcela Miozzo and Damian Grimshaw explore the relevance of theories of modularity for understanding inter-organizational collaboration and draw upon empirical evidence of IT outsourcing in Germany and the UK to support their arguments. And Chapter 5, by Volker Mahnke, Mikkel Lucas Overby and Serden Özcan, argues for the need to combine transaction cost and knowledge-based approaches to explain firm boundaries with a greater appreciation of firm characteristics, drawing on evidence of first-movers and late entrants in the US express delivery services sector.

All four chapters in Part I of the book argue for an eclectic approach to understanding inter-organizational collaboration. This means rejecting certain claims of the transaction cost approach (that firms integrate more when uncertainty increases), but perhaps rescuing other features (such as the risk of 'hold-up' generated by investment in dedicated assets). It also means drawing upon a wider literature on competences and learning in order to understand the conditions that strengthen capabilities to manage – what Nooteboom calls in Chapter 2 – competence and governance risks. When applied to the case of contracting for knowledge intensive business services, the case for an inter-disciplinary approach is strengthened further in these four chapters. Long-term, multiple and parallel relationships between clients and suppliers need to be theorized in relation to a firm's positioning itself around a future 'knowledge base' (as Howells argues in Chapter 3), to account for idiosyncratic investments in the inter-organizational coordination and transfer of knowledge (as Miozzo and Grimshaw argue in Chapter 4) and to recognize different incentive matrices facing first-movers and late entrants in KIBS markets (as Mahnke, Overby and Özcan argue in Chapter 5).

The second part of the book, 'Knowledge intensive business services in diverse national contexts', explores how the economic context and institutions of different countries shape the development of markets for knowledge intensive business services. In Chapter 6, Damian Grimshaw and Marcela Miozzo explore institutional effects on the market for IT outsourcing in Germany and the UK and seek to contribute to varieties of capitalism debates by highlighting heterogeneous within-country effects, as well as common inter-country trends involving dominant multinational IT firms. Chapter 7, by Mark Lehrer, addresses the puzzling question of how Germany went seemingly overnight from a country that preferred in-house control of IT to becoming a leading country in IT services and outsourcing. A large part of the explanation, Lehrer argues, is the success of ERP system development in a German context in which firm processes were more standardized and more explicitly codified than in other countries. In Chapter 8, Steven Casper and Sigurt Vitols show how features of Germany's coordinated market economy and the UK's liberal market economy are associated with different types of firm specialization within the computer software sector. Chapter 9 is an examination by Glenn Morgan, Andrew Sturdy and Sigrid Quack of the limitations to the growth of global firms in management consultancy. They identify the way organizational structure, power and processes impact on the internationalizing strategies of firms and the scope for autonomy exercised by subsidiaries in different national contexts.

The theme developed in Part II is that the institutional context matters in explaining how markets for knowledge intensive business services develop. The empirical bias towards comparing Germany with the UK reflects a theoretical concern to understand how markedly different institutional contexts influence economic activity, but also the need to respond to claims that the slow, incremental process of change characteristic of Germany's 'deliberative' institutions conflicts with fast-changing organizational structures ostensibly required for growth in markets for knowledge intensive business services. The four chapters demonstrate the inaccuracy of such claims, and illuminate the complex interaction between firm strategy (national and global), modes of contracting for business services and characteristics of country institutions.

A SYNTHESIS OF ARGUMENTS AND FINDINGS

The chapters in this book shed light on several sectors of knowledge intensive business services, including computer services (IT outsourcing and software applications), management consultancy, R&D services and express delivery services. Firms in these sectors typically depend on durable socio-economic relations with client organizations, suggesting that inter-organizational contracting arrangements play a key role in shaping business strategies (for both KIBS firms and client organizations) and organizational performance. But differences in the wider institutional arrangements (especially ownership relations, financial system, legal rules, labour relations and contracting standards), variation in the policies and practices of multinationals in these sectors, and the path dependency of national systems of innovation cast doubt on the emergence of a single, universal (inter-)organizational model for knowledge intensive business services. In the following, we assess the contributions of the book to these issues. We start with a review of what the chapters contribute to an interpretation of complex organizational forms found in sectors of knowledge intensive business services.

KIBS and New Organizational Forms: Consequences for Organizational Performance

Since the 1990s there have been considerable efforts at constructing a theoretical framework to understand the differential performance of firms in fast changing global markets. Despite important contributions in transaction costs and agency theory, many observers remain unconvinced by the existing economic analyses. A number of scholars working on related strands of literature, especially the management of innovation and strategic management, have attempted to fill this gap. These contributions build on insights from business history by Alfred D. Chandler, contemporary observations on business behaviour by such analysts as C.K. Prahalad and Gary Hamel, insights by behavioural analysts such as Herbert Simon, contributions by evolutionary economists such as Richard Nelson and Sydney Winter and contributions on the growth of the firm by Edith Penrose.

The emerging field on capabilities theories of economic organization explains competitive behaviour by the specialization of knowledge assets within firms (Penrose 1959) and their dynamic reconfiguration though internal, market or collaborative means. Firms are regarded as involved in incremental learning processes, through the accumulation of partly tacit knowledge, with routinized behaviour allowing firms to cope with complexity and uncertainty (Nelson and Winter 1982). In this context, firms develop dynamic capabilities which encompass the ability to build, integrate and reconfigure internal and external competences to address rapidly changing environments (Teece et al. 1997). Much productive knowledge emerges out of the interaction of many agents. The development of these capabilities therefore influences the boundaries of firms and makes it difficult to fully separate technology, production and organization.

Innovation studies has not addressed satisfactorily the broader problem of understanding how knowledge and the boundaries of the firm interact. Also, the issues of power and dependency within configurations of provision and innovation have been neglected by the innovation literature (Coombs et al. 2004). More generally, innovation studies have paid insufficient attention to the nature of inter-organizational relations, including the mechanisms through which economic coordination is achieved, competition is organized and regulated at different levels, and how rival arrangements compare and how this may influence the patterns of provision of goods and services and innovation. Furthermore, innovation studies has undertaken very little analysis of the actual patterns of economic returns among the agents involved with innovation distributed across configurations.

In this book, Chapters 2, 3 and 4 stress that the growth and differentiation among KIBS has led to the emergence of new organizational forms and that these new relationships between firms strongly influence organizational performance. The rise in KIBS cannot be simply regarded as the result of 'make-or-buy' or 'in-house-versus-outsourcing' decisions. Rather, it is a result of the increasing need to combine capabilities distributed among business units within and between firms (Coombs and Metcalfe 2000). Indeed, Jeremy Howells in Chapter 3 illustrates the complex organizational forms involved in the sourcing of innovation in the UK (contract research services and technology organizations, design and engineering consultancies and public research establishments). This process involves interaction with a range of KIBS firms, the services of which are often co-consumed together. They include much longer-term relations centred around building research and knowledge capabilities that the firm needs in the future, as well as more immediate R&D and technical requirements. Howells shows that sourcing for innovation involves the management of multiple network relations and argues that a critical business strategy is the positioning of the firm within these networks.

This complexity in the relations with KIBS is also illustrated by Marcela Miozzo and Damian Grimshaw in Chapter 4 who argue that IT outsourcing is not an all-or-nothing decision. Even when outsourcing their IT operations, client firms must still engage in active management at both the contract and relationship level, they generally tend to retain in-house capabilities and skills (the 'retained IT organization') and they must manage the transfer of staff to the IT supplier (to exchange the required industry- and firm-specific expertise in IT systems, as well as knowledge of client's business processes). Given the ubiquitous penetration of IT in all business functions and the inseparability of information from production technology, IT outsourcing also tends to be accompanied by significant changes in the wider production technologies of the client organization.

Similarly, Volker Mahnke, Mikkel Lucas Overby and Serden Özcan in Chapter 5 show that whether companies choose to develop their competences in-house or through outsourcing to KIBS is important for creating and sustaining competitive advantage. Their analysis of outsourcing in the US express delivery services during the adoption of IT-enabled innovation reveals systematic differences in the decisions (to outsource to specialist KIBS suppliers) between first- and late-movers in the express delivery sector along three dimensions: adoption risks, supplier competence and transaction risk.

The chapters use a number of different theoretical perspectives to analyse firms' decision to externalize their operations and contract with KIBS. In particular, there are differences in the extent to which transaction cost theory is viewed as useful for the analysis of firm boundaries by the different contributions. According to transaction cost economics, under more uncertainty, asset specificity and transaction frequency, firms integrate more, since inside the organization control of 'hold-up' risk (dedicated investments yield switching costs, which create dependence) is more feasible than in outside contracting (Williamson 1985). There is deverticalization and externalization of business functions under the opposite condititions. However, the complexity of the new organizational forms accompanying the rise of KIBS is difficult to deal with within the constraints of this formal theorizing.

Volker Mahnke, Mikkel Lucas Overby and Serden Özcan in Chapter 5 base their analysis on transaction cost economics, complemented with knowledge-based views, to explain changing firm boundaries. They argue that late-movers in the US express delivery industry are more likely to outsource relative to first-movers because they face supplier markets that exhibit greater relative competence and higher competition between suppliers. The other contributions in the book are more critical of transaction cost economics as a tool to explain firm boundaries. Bart Nooteboom in Chapter 2 shows that a transaction cost perspective and an approach centred on dynamic efficiency, learning and innovation yield very different conclusions regarding the decision to outsource KIBS. Nooteboom's contribution incorporates elements from different theories – the competence view of organizations and extracts from theories of knowledge – which suggest the need for complementary cognition from outside partners. The argument is that, contrary to the prediction of transaction cost economics, when uncertainty increases (that is, when technologies and markets become more complex and change faster), firms have a greater need for outside complementary competences to induce flexibility and learning. As a result, under higher uncertainty firms should use outside suppliers more rather than less. Thus, although diversity and cognitive distance might carry costs and risks, they also bring opportunities for learning.

An important point of contention is whether externalization of business functions to KIBS helps or hinders the retention or development of future core competences in client organizations. Although we may argue that the externalized functions are at first glance 'non-core', Teece (1986) proposed that the appropriation of returns on core competences might require access to complementary assets. Even if some competences are not part of core competences, they may have to be integrated in the firm. Nooteboom in Chapter 2 calls this 'competence risk' and argues that there are ways to mitigate these problems, especially by maintaining sufficient R&D in the outsourced activity to maintain 'absorptive capacity', that is, the ability to judge developments in the field (Granstrand et al. 1997). Miozzo and Grimshaw in Chapter 4 seem more doubtful about the possibility for client firms to maintain this 'absorptive capacity' in client firms involved in IT outsourcing. While a 'retained IT organization' may have the role of identifying potential future opportunities, because of the conflict of objectives between client and supplier it may be difficult to realize the required investments in innovation.

Chapters 4 and 7 analyse developments in outsourcing and the rise in KIBS on the basis of the literature on modularity. This literature analyses the link between product design and manufacturing (D'Adderio 2001; Sturgeon 2002; Zuboff 1988). It has been argued that with advances in information technology, technology convergence in application software has enabled vertical disintegration of KIBS (Pavitt 2003). The literature on complex products and systems has pointed out an important limitation of the literature on modularity – that is, the adoption of a modular product architecture does not automatically lead to a modular organization pattern. Knowledge boundaries may be different from production boundaries defined by make-or-buy decisions, and knowledge and organizational coordination demands interactive management, in many cases through 'system integrator' firms which 'know more than they do' (Brusoni and Prencipe 2001). Marcela Miozzo and Damian Grimshaw in Chapter 4 draw on empirical evidence on IT outsourcing in Germany and the UK to identify an additional limitation of the modularity literature. Modularity is often presented as a design strategy that stimulates innovation (Baldwin and

Clark 1997). The authors challenge the generalizability of this claim regarding a strong association of modularity and innovation when examining KIBS outsourcing. They argue that the intangibility of services, as determined by particular features of services such as asymmetric information between client and supplier, product/service differentiation and dynamic scale economies, exacerbates the conflicts between clients and suppliers, which may present obstacles to innovation in a strategy of modularization.

Mark Lehrer in Chapter 7 examines the exceptional changes in IT outsourcing patterns in the German IT sector. Lehrer explores different dimensions of modularity, especially market-organizational and function-organizational modularity. He suggests it is the technical complexity surrounding ERP systems that constitutes the primary causal link between function-organizational modularity (the nature of the R/3 software product) and market-organizational modularity (the delegation of ERP system development to specialized IT service firms).The complexity of ERP systems entailed market organization of both the basic software product (SAP, PeopleSoft, Baan) and of its implementation (by IT service providers). SAP benefited from a domestic market in which firm processes were more standardized and more explicitly codified than in other countries. Enterprise-wide ERP systems emerged first in Germany, Lehrer argues, because it was easier to standardize business software within the German context. The diffusion of R/2 and R/3 hastened the trend towards IT outsourcing because of the cost and technical complexity of implementation, implementation expertise accumulated more economically in outside IT firms.

Influence of a Country's Institutional Arrangements on the Evolution of KIBS

Little is known about the way in which different socio-economic contexts, as represented by national institutions and national models of economic development, shape the development of markets for KIBS. There have been several studies by geographers on the spatial organization of business services firms, with contrasting findings, for example, between studies that show most client organizations use local offices of management consultants (Wood 1998) and others where proximity is typically not very important in choice of consultancy, although more so in choice of a firm to provide software (Glückler 2000). But in his wide-ranging review of studies on KIBS, Miles (2003) calls for greater attention to the role of the wider national systems of innovation (and we would extend this to encompass business and employment) in order to understand more clearly the interactions between KIBS firms and their clients. The point that is repeated in several chapters of this book is that national institutions do play a significant role in shaping the nature and form of markets for KIBS, encompassing market growth for KIBS firms, the type of inter-organizational relations between KIBS firm and client, and the performance implications for client organizations.

In Chapters 6, 7 and 8, the theoretical point of departure is the varieties of capitalism approach (Hall and Soskice 2001; Whitley 1999). This approach argues that national patterns of specialization are created (and reinforced) by comparative institutional advantages in managing particular organizational competences. A growing number of studies show that such competences include those related to innovation (Casper et al. 1999), human capital investment (Finegold and Soskice 1988; Lazonick and O'Sullivan 1996), strategic management (Lehrer 2001) and production organization (Streeck 1992). Applying this approach to a country's specialization in KIBS reveals a number of findings that extend a varieties of capitalism approach in several ways.

The first point is that while the varieties of capitalism approach effectively illuminates the combinative nature of institutions, with the possibility of different 'coherent' models promoting different paths of economic growth, it also perhaps inadvertently encourages stylized interpretations of liberal market economies as more open to new markets, new products and radical innovation than the coordinated market economies that are hampered by deliberative, slow-moving institutions. Empirical analysis of KIBS sectors is an ideal focus to test such a proposition, since they typically involve new technologies and new markets for services provision.

What the evidence collected in Part II of this book shows is that certain KIBS sectors have in fact performed remarkably well in the archetypal coordinated market economy, Germany. In the case of the market for software applications, Mark Lehrer in Chapter 7 argues that the specific combination of institutions in Germany established a unique foundation that propelled market growth. Two institutional features were crucial: an 'employment effect' (generated by the institutions of training, trade union organization, labour law and pension plans that favour long-term and stable employment relations); and a 'standardization effect' (caused by the high coordination of industry and labour organizations that generates strong homogeneity of business practices). In Mark Lehrer's words, 'ERP systems emerged first in Germany because it was easier to standardize business software within the German context'. The case of the IT outsourcing market is similar. In Chapter 6, Damian Grimshaw and Marcela Miozzo show that Germany experienced rapid growth in the IT outsourcing market, to a level comparable to that found in the UK, despite contrasting institutional environments. Negotiating IT outsourcing contracts did involve greater deliberation in Germany than in the UK, especially with works councils. But the main consequence was not the preservation of the status quo, but rather the use of alternative organizational forms (namely, joint ventures between clients and IT firms) which would ensure a smooth transition of IT professionals to an external services provider. The result, arguably, was to establish IT outsourcing arrangements on a more stable footing than those found in the UK where the risk of industrial relations problems and misunderstanding of contractual commitments between client and supplier was higher. In both chapters, the argument is that diverse institutional contexts are associated with diverse paths of growth in KIBS markets, not that one combination of institutions favours KIBS growth more than another.

A second theme is that a national model need not necessarily imply a uniform institutional structure across the economy. For example, the recent expansion of equity markets in Germany, while small, breaks the uniformity of its bank-centred financial system (Lehrer 2000). Changes in the 'hierarchy' (Amable 2000) of these institutional arrangements may shape the development of different industries. In Chapter 8, for example, Steven Casper and Sigurt Vitols find evidence that a large group of German software firms have adopted capital-market-based financing methods – evidence of significant change in the German financial system – although this appears to prevail alongside 'traditional' features of a high concentration of shares owned by high-level management.

A third theme is that strongly internationalized sectors, as found in many areas of KIBS, present different challenges, and pose different questions, for a varieties of capitalism approach. Instead of exploring how alternative institutional environments encourage different national patterns of specialization, the enquiry shifts to their degree of openness to foreign direct investment, adaptability to international standardized business practices and ability to establish productive linkages with foreign-owned multinationals. It also means assessing how national 'coherent' models are confronted by the 'challenger rules' of powerful multinational firms (Djelic and Quack 2003).

Damian Grimshaw and Marcela Miozzo show in Chapter 6 how acquisition of German IT firms by foreign-owned multinationals led to a standardization of organizational forms with a buy-out of joint ventures over a relatively short period and a shift to direct outsourcing of IT services. The neat match between institutions and joint venture forms in Germany thus appears to have been challenged by multinational firms' goals to establish standardized business practice across countries. Nevertheless, the idea that multinationals necessarily enjoy distinctive advantages due to their global organization is questioned in Chapter 9 by Glenn Morgan, Andy Sturdy and Sigrid Quack. Their study of global consultancy firms shows that a significant part of business activity depends on the relatively autonomous decisions and strategies of subsidiaries in their bid to win business and distribute rewards. Consulting firms are global, they argue, because this underpins reputation and legitimacy (needed for recruitment and selection, up or out promotion, and attracting prestigious clients) and facilitates economies of scale, scope, repetition and learning in services provision. But the broader career environment (especially the 'tournament' for promotion and the billable hours system) reduces the willingness of consultants to share knowledge and frees up little time to participate in global knowledge management teams. These political and organizational factors thus conspire against global coordination and, because of the high level of costs and high coordination requirements of such firms, their services are limited to large-scale projects and clients. Beneath this level, there is an opportunity for other forms of management consulting organizations, local and across national borders.

We conclude our introduction to this book with the disclaimer that the coverage of a book of this nature is inevitably selective. However, we are confident that the contributions in this book offer original analysis and an interesting collection of empirical findings that can stimulate further work on KIBS, organizational forms and the influence of countries' institutional arrangements.

NOTES

1. Among the larger OECD economies, services production growth was double, and sometimes triple, the growth of manufacturing. The one exception was Sweden where, during 1993–2001, manufacturing production grew at almost twice the rate of services (own calculations, STAN database, OECD 2003).
2. Our approximate and incomplete definition includes the following sectors of activity: ISIC code 72 (hardware consultancy, software consultancy, data processing, database activities, maintenance and repair of office and computing machinery and other computer related activities); ISIC code 73 (R&D on natural sciences and engineering and R&D on social sciences and humanities); and ISIC code 74 (legal, accounting and auditing activities, market research, management consultancy, architectural engineering, advertising and other business activities). The main weakness of this definition of KIBS is that one component of code 74, namely 749 'business activities n.e.c.', includes several non-KIBS activities; 749 includes labour recruitment and provision of personnel, investigation and security services, building and office cleaning, photographic and packaging activities. However, STAN data are only provided at the two-digit level of industry classification.
3. It is notable that business services also far outstripped other sectors in the generation of jobs for low-educated workers (EC 2003, Figure E). This is because this broader category includes 'knowledge intensive business services' as well as 'operational services', such as industrial cleaning and security.
4. Thanks are due to Professor Rod Coombs (then one of the Co-Directors of CRIC) for collaborating in discussions outlining the workshop design, as well as Professor Richard Whitley for his help in chairing some of the sessions.

REFERENCES

Amable, B. (2000), 'Institutional complementarity and diversity of social systems of innovation and production', *Review of International Political Economy*, **7** (4), 645–87.

Antonelli, C. (1999), *The Micro Dynamics of Technological Change*, London: Routledge.

Arora, A. and A. Gambardella (1994), 'The changing technology of technical change: general and abstract knowledge and the division of innovative labour', *Research Policy*, **23** (5), 523–32.

Baldwin, C.Y. and K.B. Clark (1997), 'Managing in an age of modularity', *Harvard Business Review*, **75** (5), 84–94.

Baumol, W. (1967), 'Macroeconomics of unbalanced growth: the anatomy of urban crisis', *American Economic Review*, **57** (3), 413–26.

Brusoni, S. and A. Prencipe (2001), 'Unpacking the black box of modularity', *Industrial and Corporate Change*, **10** (1), 179–205.

Casper. S., M. Lehrer and D. Soskice (1999), 'Can high-technology industries prosper in Germany? Institutional frameworks and the evolution of the German software and biotechnology industries', *Industry and Innovation*, **6** (1), 5–24.

Cohendet, P. and F. Meyer-Krahmer (2001), 'Editorial', *Research Policy*, **30** (9), 1353–4.

Coombs, R. (1999) 'Innovation in services: concerning the services–manufacturing divide', Nijmegen Lectures on Innovation Management, Nijmegen Business School, Antwerpen-Apeldoorn, Maklu.

Coombs, R. and S. Metcalfe (2000), 'Organizing for innovation: co-ordinating distributed innovation capabilities', in N. Foss and V. Mahnke (eds), *Competence, Governance, and Entrepreneurship: Advances in Economic Strategy Research*, Oxford: Oxford University Press, pp. 209–31.

Coombs, R., M. Harvey and B. Tether (2004), 'Analysing distributed processes of provision and innovation', *Industrial and Corporate Change*, **12** (6), 1125–55.

D'Adderio, L. (2001), 'Crafting the virtual prototype: how firms integrate knowledge and capabilities across organisational boundaries', *Research Policy*, **30** (9), 1409–24.

David, P. and D. Foray (2003), 'Economic fundamentals of the knowledge society', *Policy Futures in Education*, e-journal, **1** (1), 20–49.

Di Cagno, D. and V. Meliciani (2005), 'Do inter-sectoral flows of services matter for productivity growth? An input–output analysis on OECD countries', *Economics of Innovation and New Technology*, **14** (3), 149–72.

Djelic, M-L. and S. Quack (2003), 'Introduction: governing globalization – bringing institutions back in', in M-L. Djelic and S. Quack (eds), *Globalization and Institutions: Redefining the Rules of the Economic Game*, Cheltenham, UK and Northampton, MA, USA: Edward Elgar, pp. 1–14.

Drejer, I. (2001), *Business Services as a Production Factor*, business studies working paper, Center for Economic and Business Research, Ministry of Economics and Business Affairs, Copenhagen.

EC (1998), communication from the Commission to the Council, 'The contribution of business services to industrial performance – a common policy framework', Brussels: EC.

EC (2003), communication from the Commission to the Council, the European Parliament, the European Economic and Social Committee and the Committee of the Regions, 'The competitiveness of business-related services and their contribution to the performance of European enterprises', COM (3002) 747, Brussels: EC.

Finegold, D. and D. Soskice (1988), 'The failure of training in Britain: analysis and prescription', *Oxford Review of Economic Policy*, **4** (3), 21–53.

Gadrey, J. and F. Gallouj (eds) (2002), *Productivity, Innovation and Knowledge in Services: New Economic and Socio-Economic Approaches*, Cheltenham, UK and Northampton, MA, USA: Edward Elgar.

Gallouj, F. and O. Weinstein (1997), 'Innovation in services', *Research Policy*, **26** (4–5), 537–56.

Glückler, J. (2000), 'Management consulting: structure and growth of a knowledge intensive business service market in Europe', Forschungsberichte working papers, ISSN 1439–2399, Institut für Wirtschafts und Sozialgeographie der Johann Wolfgang Goethe, Universität Frankfurt.

Grandstand, O., P. Patel and K. Pavitt (1997), 'Multi-technology corporations: why they have "distributed" rather than "distinctive core" competences', *California Management Review*, **39** (4), 8–25.

Greenhalgh, C. and M. Gregory (2000), 'Labour productivity and product quality: their growth and inter-industry transmission in the UK 1979–1990', in R. Barrell, G. Mason and M. O'Mahony (eds), *Productivity, Innovation and Economic Performance*, Cambridge: Cambridge University Press, pp. 58–92.

Hall, P. and D. Soskice (eds) (2001), *Varieties of Capitalism: The Institutional Foundations of Comparative Advantage*, Oxford: Oxford University Press.

Kasoulacos, Y. and N. Tsounis (2000), 'Knowledge-intensive business services and productivity growth: the Greek evidence', in M. Boden and I. Miles (eds), *Services and the Knowledge-based Economy*, London and New York: Continuum, pp. 192–208.

Lazonick, W. and M. O'Sullivan (1996), 'Organisation, finance and international competition', *Industrial and Corporate Change*, **5** (1), 1–49.

Lehrer, M. (2000), 'Has Germany finally solved its high-tech problem? The recent boom in German technology-based entrepreneurship', *California Management Review*, **42** (4), 89–107.

Lehrer, M. (2001), 'Macro-varieties of capitalism and micro-varieties of strategic management in European airlines', in P. Hall and D. Soskice (eds), *Varieties of Capitalism: The Institutional Foundations of Comparative Advantage*, Oxford: Oxford University Press, pp. 361–86.

Miles, I. (1999), 'Foresight and services: closing the gap?', *Service Industries Journal*, **19** (2), 1–27.

Miles, I. (2001a), 'Knowledge-intensive business services and the New economy', paper presented at the Evolutionary Economics Unit, Max-Planck Institute for Research into Economic Systems, September.

Miles, I. (2001b), *Knowledge Intensive Business Services Revisited*, Nijmegen Lectures on Innovation Management, Antwerpen-Apeldoorn, Maklu.

Miles, I. (2003), 'Business services and their contribution to their clients' performance: a review', ECORYS/CRIC project, University of Manchester, mimeo.

Miozzo, M. and I. Miles (2002), *Internationalization, Technology and Services*, Cheltenham, UK and Northampton, MA, USA: Edward Elgar.

Miozzo, M. and L. Soete, (2001), 'Internationalization of services: a technological perspective', *Technological Forecasting and Social Change*, **67** (2), 159–85.

Nelson, R. and S. Winter (1982), *An Evolutionary Theory of Economic Change*, Cambridge, MA: Harvard University Press.

Organization for Economic Co-operation and Development (OECD) (1999), *Strategic Business Services*, Paris: OECD.

OECD (2003), *STAN Structural Analysis Database*, Paris: OECD.

OECD (2005), *Enhancing the Performance of the Services Sector*, Paris: OECD.

Pavitt, K. (2003), 'What are advances in knowledge doing to the large industrial firm in the "New Economy"?', in J.F. Christensen and P. Maskell (eds), *The Industrial Dynamics of the New Digital Economy*, Cheltenham, UK and Northampton, MA, USA: Edward Elgar, pp. 103–20.

Peneder, M., S. Kaniovski and B. Dachs (2001), 'External services, structural change and industrial performance', Austrian Institute of Economic Research (WIFO), background report for The European Competitiveness Report 2000, Enterprise DG working paper no 3, Brussels: EC DG-Entreprise.

Peneder, M., S. Kaniovski and B. Dachs (2003), 'What follows tertiarisation? Structural change and the role of knowledge-based services', *Services Industries Journal*, **23** (2), 47–66.

Penrose, E. (1959), *The Theory of the Growth of the Firm*, Oxford: Oxford University Press.

Petit, P. (2002) 'Growth and productivity in a knowledge-based service economy', in J. Gadrey and F. Gallouj (eds), *Productivity, Innovation and Knowledge in Services: New Economic and Socio-Economic Approaches*, Cheltenham, UK and Northampton, MA, USA: Edward Elgar, pp. 102–23.

Reich, R. (1990), *The Work of Nations: Preparing Ourselves for 21st-Century Capitalism*, New York: A.A. Knopf.

Sapir, A. (1987), 'International trade in services: comments', in O. Giarini (ed.), *The Emerging Service Economy*, Oxford: Pergamon Press.

Stigler, G.J. (1951), 'The division of labour is limited by the extent of the market', *Journal of Political Economy*, **59** (3), 185–93.

Streeck, W. (1992), *Social Institutions and Economic Performance: Studies of Industrial Relations in Advanced Capitalist Countries*, London: Sage.

Sturgeon, T.J. (2002), 'Modular production networks: a new American model of industrial organization', *Industrial and Corporate Change*, **11** (3), 451–96.

Teece, D.J. (1986), 'Profiting from technological innovation: implications for integration, collaboration, licensing and public policy', *Research Policy*, **15** (6), 285–35.

Teece, D.J., G. Pisano and A. Shuen (1997), 'Dynamic capabilities and strategic management', *Strategic Management Journal*, **18** (7), 509–33.

Tether, B.S. and G.M.P. Swan (2003), 'Services innovation and the science base: an investigation into the UK's system of innovation using evidence from the UK's third community innovation survey', presented at the Institute of Socio-Economic Studies on Innovation and Research Policy, National Research Council, ISPRI-CNR and University of Urbino, Faculty of Economics, International Workshop on Innovation in Europe: Empirical Studies on Innovation Surveys and Economic Performance, Rome 28, January 2003.

Tomlinson, M. (2000), 'Information and technology flows from the service sector: a UK–Japan comparison', in M. Boden and I. Miles (eds), *Services and Knowledge-Based Economy*, London and New York: Continuum, pp. 209–21.

Whitley, R. (1999), *Divergent Capitalisms: The Social Structuring and Change of Business Systems*, Oxford: Oxford University Press.

Williamson, O.E. (1985), *The Economic Institutions of Capitalism*, New York: Free Press.

Wood P. (1998), 'Services and internationalization', report to the RESER network, accessed 10 October, 2003 at www.reser.net/gb/rapportgb.html.

Zuboff, S. (1988), *In the Age of the Smart Machine: The Future of Work and Power*, Oxford: Heinemann.

APPENDIX

Table A.1.1   *Average annual real growth in KIBS, services and manufacturing production, 1991–2001*

	ISIC rev 3	Austria 1991–01 (%)	France 1991–01 (%)	Germany 1991–01 (%)	Italy 1992–01 (%)	Netherlands 1991–01 (%)	Spain 1995–99 (%)	Sweden 1993–01 (%)	UK 1992–00 (%)
Total KIBS sectors	72–74	7.2	4.0	4.0	5.9	6.4	8.1	10.2	9.0
Computer and related activities	72	14.3	7.1	8.4	5.2	13.0	14.5	16.2	13.1
Research development	73	8.7	1.3	3.3	n.a.	2.8	3.1	n.a.	2.8
Other business activities	74	5.9	3.9	3.4	6.2	6.4	7.2	8.5	8.4
Services	50–99	3.2	2.5	2.0	2.6	4.1	4.0	3.7	5.0
Manufacturing	15–37	1.6	1.5	0.3	1.7	1.8	2.9	6.4	1.1
Total economy	01–99	2.5	1.9	1.0	2.0	3.1	3.2	4.2	3.5

Notes: Current price production data from the STAN database adjusted using OECD estimates of Consumer Price Index for each country. Own calculations.
Detailed STAN data for sectors 72–74 are missing for the USA, Japan, Canada and Australia.

Source: OECD STAN Structural Analysis Database (2003) and OECD Economic Outlook (2003).

Table A.1.2 Production in KIBS, manufacturing and services, as a share of total GDP, 1991–2001

	ISIC rev 3	Netherlands 1991 %	Netherlands 2001 %	Spain 1995 %	Spain 1999 %	Sweden 1993 %	Sweden 2001 %	UK 1992 %	UK 2000 %	Austria 1991 %	Austria 2001 %	France 1991 %	France 2001 %	Germany 1991 %	Germany 2001 %	Italy 1992 %	Italy 2001 %
Total KIBS sectors	72–74	6.5	9.4	4.0	4.8	6.4	10.1	6.6	10.0	3.9	6.1	9.1	11.3	6.5	8.8	5.8	8.1
Computer and related activities	72	0.6	1.5	0.4	0.7	1.2	2.8	1.1	2.1	0.4	1.3	1.2	2.0	0.7	1.4	1.1	1.5
Research and development	73	0.4	0.4	n.a.	n.a.	n.a.	n.a.	0.4	0.3	0.1	0.2	1.2	1.2	0.3	0.4	n.a.	n.a.
Other business activities	74	5.4	7.4	3.6	4.1	6.1	8.5	5.2	7.6	3.5	4.8	6.7	8.1	5.5	7.0	4.7	6.6
Services	50–99	52.3	58.0	50.6	52.1	61.1	58.7	57.7	65.0	53.9	58.0	54.3	57.7	50.8	55.8	53.8	56.9
Manufacturing	15–37	30.5	26.7	32.4	32.0	28.6	33.3	27.8	23.0	31.6	29.0	38.3	40.9	37.9	34.9	33.9	32.7
Total economy	01–99	100.0	100.0	100.0	100.0	100.0	100.0	100.0	100.0	100.0	100.0	100.0	100.0	100.0	100.0	100.0	100.0

Source: OECD STAN Structural Analysis Database (2003).

25

PART I

Knowledge intensive business services
and organizational forms

2. Principles of inter-organizational relationships: an integrated survey

Bart Nooteboom

INTRODUCTION

The sourcing decision – what to make and what to buy – is a special case of the more general decision of what to do inside one's own organization, and what to do outside, in collaboration with other organizations. Outsourcing entails vertical collaboration in the supply chain. Other forms of collaboration may be horizontal, with competitors, or lateral, with firms in other industries. Next to the question *what* to do inside or outside, and *why*, there are the questions *with whom* to collaborate, and *how*, in what *forms of organization* and with what *instruments for governance*.

The time is ripe for a unified approach that analyses all these questions, integrating different perspectives and arguments that have emerged in recent advances, from different disciplines. In particular, there is a need to combine perspectives of competence and governance (Nooteboom 2004a; Williamson 1999). From the perspective of 'dynamic competences', there has been a focus on the development of competences and learning, with a neglect of relational risk and its governance. Transaction cost economics (TCE) has focused on 'hold-up risk' and the hazards of opportunism, to the neglect of learning and innovation. Concerning governance, network analysis in the social sciences has included the role of trust, while trust has been neglected, or even ruled out, in TCE. In a wide literature, in economics, sociology and business, important new insights have been generated, but they tend to focus on few aspects, resulting in one-sided conclusions. For example, concerning sourcing, opinion seems to have settled on a rather extreme view in favour of outsourcing everything that is not part of 'core competences'. This is problematic (Bettis et al. 1992), as will be discussed.

This chapter looks at relations not only between firms, but between organizations more in general, which may not sell products in markets, and may not be independent legal entities. This includes many organizations in the (semi-)public sector. This chapter looks not only at the positive but also at the negative side of relationships, and at how relationships may be ended.

Outside sociology there still is a tendency to look at relations in terms of dyads. This chapter also looks at network effects of multiple partners and indirect linkages, studied in sociology (Burt 1992; Granovetter 1973; Powell 1990; Uzzi 1996). This chapter avoids claims of universal best practice. Forms of organization and governance are contingent upon a range of conditions.

The chapter proceeds as follows. First comes an integrated summary of theoretical perspectives concerning competence and governance. Then there is a summary of goals of collaboration, forms of organization and the choice between them, instruments for governance of relational risk, contingences for their choice, network effects, and relationship development. Extensive use is made of Nooteboom (2004b), which gives a more detailed discussion.

THEORETICAL PERSPECTIVES

Competence

The perspective of TCE yields useful insights, but is also subject to serious limitations. First, by its own admission (Williamson 1985, 1999), it has little to say about learning and innovation. Second, contrary to what Williamson (1993) claimed, trust and loyalty matter. Theoretical sources beyond TCE, in sociology and cognitive science, are needed to incorporate learning and trust (Nooteboom 1992). The claim of this chapter is that these elements from different theories can be integrated into a coherent whole.

A second perspective draws from the 'competence view' of organizations, which goes back to the work of Edith Penrose (1959). While transaction cost economics focuses on static efficiency – by trading off production costs, transaction costs and costs of organization, given a certain state of knowledge, technology and preferences – we require a perspective of dynamic efficiency or innovation, incorporating shifts of knowledge, technology and preferences. It is now a priority for firms to develop 'dynamic capabilities'. For this, they need to maintain flexibility of configurations of competences, for the sake of innovation in the form of Schumpeterian 'novel combinations'. This yields the claim that firms should concentrate on the activities at which they are best, and outsource other activities as much as strategically possible. For example, in order to reduce development times of new products and to reduce risks of maladjustment to customer needs, the supplier should be brought in as a partner in developing and launching a new product. However, note the qualification of strategic possibilities. One may need to integrate activities in order to control risks of dependence and spillover, or to preserve options for future core competences.

In addition to the usual considerations of efficiency, flexibility and speed, learning is an important goal of collaboration. This is supported by a theory of knowledge that suggests that people's perceptions, interpretations and value judgements are dependent on mental frameworks (or schemata, or categories) that develop from the experience they have in interaction with their (physical and social) environment. This view goes back to the work of Piaget and Vygotsky, in developmental psychology, and the 'symbolic interactionism' of G.H Mead, in sociology, and has later been called the 'experiential' view of knowledge (Kolb 1984) and the 'activity' view (Blackler 1995). The resulting mental frameworks constitute 'absorptive capacity' (Cohen and Levinthal 1990). Such cognitive construction implies that people who have constructed their cognition in different conditions, along different life histories, will see, interpret and evaluate the world differently.

This difference yields an opportunity for innovation, but also a problem of organization, in the coordination of knowledge and norms of behaviour. In organizations, a cognitive focus is needed of shared perceptions, interpretations and values, for sufficient mutual absorptive capacity to link different competencies and to align objectives and motives. Such a focus includes elements of cognition in the narrow sense of understanding the world (competence) and elements concerning ways in which people deal with each other (governance). This yields the idea of an organization as a 'sensemaking system' (Weick 1979, 1995), 'system of shared meaning' (Smircich 1983), 'focusing device' (Nooteboom 1999), or 'interpretation system' (Choo 1998). This is more fundamental for organizations than the need to reduce transaction costs.

However, such organizational focus creates a risk of myopia, which needs to be redressed by employing complementary cognition from outside partners, at a 'cognitive distance' that is sufficiently large to yield novel insight and sufficiently small to ensure that it is still comprehensible (Nooteboom 1992, 2000, 2004b; Wuyts et al. 2005).

This yields a prediction that is an opposite to a prediction from TCE. According to the latter, firms integrate more when uncertainty increases, since inside the organization control of 'hold-up' risk is more feasible than in outside contracting. Here, the argument is that when uncertainty increases, in the sense that technologies and markets become more complex and change faster, firms have a greater need for outside complementary competence, for the sake of flexibility and learning. As a result, under higher uncertainty firms should use outside suppliers more rather than less. The hypothesis of an increased need of alliances under conditions of volatility has been confirmed by Colombo and Garrone (1998). They found that in technologically volatile industries, as measured by patent intensity, the likelihood of alliances is higher than in the absence of such volatility.

Governance

Competences are not off-the-shelf products but are embedded in the heads and hands of people, in teams, organizational structure and procedures, and organizational culture. They have a strong tacit dimension, especially in innovation. Their development is path-dependent: they build upon pre-existing firm-specific assets and organizational learning (Lippman and Rumelt 1982). Ongoing or intermittent interaction is needed to enable the exchange of tacit knowledge. Some scholars conclude that this requires full organizational integration, but that is denied here. However, the exchange and joint production of knowledge between firms with different perspectives and competences does require mutual absorptive capacity and a shared language for communication, to cross 'cognitive distance'. This takes time to develop, and can require a dedicated investment, so that relations have to last a sufficiently long time to make the investment worthwhile. Here, in this new perspective on specific investments, needed to achieve mutual understanding, there is a connection between the competence perspective and transaction cost thinking.

Mutual understanding and trust emerge in a process of interaction. As a result, we need to consider not transactions by themselves, as in TCE, but transactions in the setting of an exchange relationship that develops in time (Granovetter 1985; Gulati 1995; Helper 1987; Ring and van de Ven 1992; Sako 1992).

According to TCE, it is impossible to reliably judge possible limits to other people's opportunism, and therefore trust does not yield a reliable safeguard (Williamson 1975, pp. 31–7). If trust goes beyond calculative self-interest, it yields blind, unconditional trust, which is not wise and will not survive in markets (Williamson 1993). From a social science perspective, many others take the view that trust is viable, without necessarily becoming blind or unconditional (Bromiley and Cummings 1992; Bradach and Eccles 1989; Chiles and McMackin 1996; Deutsch 1973; Dyer and Ouchi 1993; Gambetta 1988; Granovetter 1973; Gulati 1995; Helper 1990; Hill 1990; Macauley 1963; McAllister 1995; Murakami and Rohlen 1992; Noorderhaven 1996; Nooteboom et al. 1997; Nooteboom 1999; Nooteboom 2002; Ouchi 1980; Ring and van de Ven 1994). Man is not solely self-interested and opportunistic, in business also common honesty and decency are found. Partners may develop mutual empathy. They may voluntarily refrain from opportunism. Trust, then, enables a leap beyond the expectations that reason and experience alone would warrant (Bradach and Eccles 1989), and thus goes beyond formal control mechanisms. A committed partner does not immediately exit from the relationship in case of unforeseen opportunities or problems, but engages in 'voice' (Helper 1987; Hirschman 1970). However, voice and commitment need not and should not entail that relations last endlessly. Indeed, relations can become too durable, and too exclusive, with too much mutual identification and trust, yielding rigidities and lack of the variety that is needed for learning. Social capital can deteriorate into social liability (Gargiulo and Benassi 1999; Leenders and Gabbay 1999).

In spite of fundamental criticism of TCE, indicated above, it still yields useful insights for governance. It has been used empirically with sometimes considerable and sometimes limited success. Its core insight is that dedicated investments yield switching costs, which create dependence, resulting in a risk of 'hold-up'. Given hazards of opportunism, this risk must be 'governed'. That insight has been widely corroborated in empirical studies (Anderson and Gatignon 1986; Berger et al. 1995; Chiles and McMackin 1996; Hennart 1988; Nooteboom et al. 1997; Parkhe 1993). TCE next claims that when the risk is high, it requires organizational integration to control it (Williamson 1975). Full integration entails merger or acquisition (MA), but there are intermediate forms of semi-integration by means of detailed 'bilateral governance' (Williamson 1985). When the frequency of transactions does not warrant the investment in detailed bilateral governance, one can employ a third party for arbitration, in 'trilateral' governance. Here, in the prediction of organizational forms of collaboration, the empirical results of TCE are mixed and inconclusive (Carter 2002). I propose that this is due to lack of attention to dynamic capabilities. As suggested above, uncertainty may lead to more, not less outside partnerships, and looser, not tighter control.

Propositions

The analysis yields the following propositions:

H1 While neglected in TCE, learning and trust are essential features of inter-firm relationships.

H2 Counter to TCE, the relationship rather the transaction should be the unit of analysis.

H3 Counter to TCE, for the existence of the firm, the creation of a cognitive focus is more fundamental than the reduction of transaction costs.

H4 Obtaining complementary cognitive scope is an additional argument for external relations.

H5   Counter to TCE, under greater uncertainty, complexity and change, firms (should) outsource more, not less, with looser, not tighter control.

H6   TCE still yields useful features concerning relational risk and instruments for governance.

H7   Relationships can deteriorate into rigidities, and ending a relationship may be as important as starting one.

GOALS AND RISKS

Goals

There are a variety of goals of collaboration. Table 2.1 gives a summary. The goals are grouped into goals of efficiency, competence and 'positional advantage' (Stoelhorst 1997). Almost all of these goals of collaboration are familiar from the literature, and will not be fully discussed here (e.g. Anderson and Gatignon 1986; Contractor and Lorange 1988; Faulkner 1995; Hennart 1988; Jarillo 1988; Killing 1983; Lamming 1993; Porter and Fuller 1986; Ohmae 1989).

Table 2.1   Goals of collaboration

Efficiency
E1   Avoid overcapacity
E2   Economy of scale, scope or time
E3   Spread risk
E4   Combine or swap products

Competences
C1   Complementary competences
C2   Variety of learning
C3   Flexibility of configuration

Positioning
P1   Satisfy demands from host government on local content, repatriation of profits, use of expatriates
P2   Fast access to new markets of products and inputs
P3   Adjustment of products, technology or inputs to local markets and conditions
P4   The offer of a joint product package
P5   Attack a competitor in his home market
P6   Establish a standard in the market
P7   A cartel

A few items may require explanation. Economies of time entail the reduction of idle time of people, installations, or goods in stock, set-up or switching. Collaboration can help, for example, by reducing stocks in just-in-time delivery. The swapping of products refers, for example, to cross-licensing, when the outcomes of R&D are unpredictable and turn out to yield products that do not fit in production or distribution (E4). An example is pharmaceutical products. Concerning competence, one may need to use complementary competences of others that one could not oneself develop fast enough (C1). A new point concerns cognitive competence, with the need to prevent organizational myopia by supplementing perspectives from others, for extending 'cognitive scope', as indicated in the previous paragraph (C2). There is also an important argument of flexibility: one can more easily reconfigure outside relations than build up and scrap activities within the firm (C3). Note that there are institutional differences here between different economies. In the USA it is easier to hire and fire and to buy and sell parts of firms than it is in Europe. As a result, the need for outside collaboration is greater in (continental) Europe than in the USA, resulting more in 'networked economies'.

Collaboration even between competitors may be desirable for developing and setting a joint standard in the market (P6), or for offering a joint port-folio of products (P4), especially in industrial markets. Large, global customers want a customized portfolio of goods/services at a range of different locations. We see this, for example, in ICT services. Of course a cartel – making price agreements or dividing the market between competitors (P7) – is forbidden by law, but firms may yet try it when they get the opportunity. The goals represent the positive side of collaboration, but there is also a downside, in risks that it may entail. One risk concerns loss of competences and the other is relational risk, which has two forms: 'hold-up' and spill over risk.

Competence risk

By outsourcing one may surrender the capability to assess the value of the offering of suppliers (Beije 1998). Another problem is that one may drop a capability that later turns out to be crucial in order to utilize or replace elements of core competence. Teece (1986) proposed that the appropriation of returns on core competences might require access to complementary assets. Even if those are not part of core competence, they may have to be integrated into the firm. One may therefore have to see such complementary assets as attached to core competence. In fact, some people argue that because of these problems outsourcing does not increase flexibility, as I argued above, but decreases it (Bettis et al. 1992; Mol 2001).

However, there are ways to mitigate these problems. One is to make use of a benchmarking service, so that one can compare a supplier's offering with best practice. A second is to maintain sufficient R&D in the outsourced activity to maintain 'absorptive capacity', i.e. the ability to judge developments in the field. This may also help to retain the option of re-entry later, to retain options for future core competencies, perhaps as a second-mover, but still fast enough to be a serious player. This is reflected in empirical evidence that firms retain an R&D capability in activities that were outsourced (Granstrand et al. 1997). However, such R&D can perhaps be done in collaboration with others, in an R&D consortium. One may also try to retain the required openings in distribution channels by means of alliances. In other words, outside collaboration may also be used to retain options for the utilization or modification of core competences. Here, the flexibility of outside collaboration returns: one may also use it to maintain more flexibility in options for future core competence.

Governance Risk

As defined in TCE, the problem of hold-up results from dependence, in the form of switching costs: one incurs a loss if the relationship breaks. Part of that is the loss of relation-specific investments, and the need for new ones in another relationship. As also argued by TCE, one may lose a 'hostage'. There are also opportunity costs: the loss of the value that the current partner offers relative to the next best alternative. This depends on the availability of alternative partners, or the possibility of conducting an activity oneself, and the extent that the partner offers something unique. In other words, it depends on the extent that a partner has a monopoly in his offering, or monopsony in his access to markets. The partner may achieve this by engaging in specific investments, to establish a unique offer. The 'hold-up' risk is that the partner may opportunistically use asymmetric dependence to demand a higher share of jointly produced added value, under the threat of exiting from the relationship.

One refinement is needed. Specificity of products does not necessarily entail specificity of the assets needed to produce them. If production technology is flexible, one can produce specific, differentiated products with generic assets. This is important, because one of the effects of ICT is that a number of processes in design, production and marketing have become more flexible, thereby reducing the problem of specific investments. For example, in production one can change the program in a computer numerically controlled (CNC) machine relatively easily, to craft different shapes and functions. Virtual specification and testing of prototypes, in computer simulation, yields much more flexibility and speed of development than real, physical prototyping and testing.

Spillover risk entails that knowledge that constitutes competitive advantage, as part of core competence, reaches competitors and is used by them for imitation and competition. The risk may be direct, in the partner becoming a competitor, or indirect, with knowledge spilling over to a competitor via a partner. In the past many firms have been overly concerned with spillover risk. First of all, one should realize that to get knowledge one must offer knowledge. The question is not how much knowledge one loses, but what the net balance is of giving and receiving knowledge. Second, when knowledge is tacit it spills over less easily than when it is documented. However, even then it can spill over, for example when the staff or the division in which the knowledge is embedded are poached, or when the staff involved have more allegiance to their profession than to the interests of the firm (Grey and Garsten 2001), and professional vanity leads them to divulge too much in meetings with outside colleagues. Furthermore, the question is not whether information reaches a competitor but whether he or she will also be able to turn it into effective competition. For this they need to understand it, and their absorptive capacity may not enable that. There may be 'causal ambiguity' (Lippman and Rumelt 1982). Next, they will need to effectively implement it in their organization. And finally, if by that time the knowledge has shifted, one does not care.

Propositions

> **H8** To some extent, more flexible technology has reduced the importance of specific investments, reducing hold-up risk. However, specific investments in developing mutual understanding and in the development of trust have become more pronounced.
>
> **H9** Spillover risk is often low, due to constraints in absorptive capacity of competitors and rapid change of knowledge.
>
> **H10** Outsourcing may yield loss of absorptive capacity and options for future core competences. However, there are also ways to maintain them in outside relations.

ORGANIZATIONAL FORMS

Mergers/Acquisitions or Alliances?

In collaboration, full integration entails a merger or acquisition (MA). Short of that, there is a wide range of possible forms of collaboration 'between market and hierarchy'. One intermediate form is a joint venture (JV), with a sharing of ownership and management among the partners. Other forms are: franchising, consortia, associations, Japanese *keiretsu* and 'industrial districts', among others. These forms differ in centralization of ownership and management, number of participants, frequency of interaction and durability and content of collaboration. Consortia are often used for research, to spread costs and risks between multiple partners. Associations are typically used for joint advertising, lobbying, training and certification. Franchising is typically used for distribution.

A central question is when and why firms should engage in full integration, in an MA, and when they should remain more or less separate, in an alliance, with an equity joint venture as an intermediate form. Table 2.2 summarizes the argument for the alternatives of an MA and an alliance. Overall, the argument for integration, in an MA, is that it yields more control, in particular of hold-up and spillover risk. For hold-up, the argument comes from TCE. Within a firm, under the sway of administrative fiat, one can demand more information for control and one can impose more decisions than one could in respect of an independent partner. A similar argument applies to spillover risk: one can monitor and control better what happens to information.

Overall, the argument for an alliance is that it allows partners to maintain more focus of core competence, more flexibility of configuration and more variety of competence for the sake of innovation and learning, as discussed before. Also, as recognized in TCE, an independent firm that is responsible for its own survival will be more motivated to perform than an internal department that is assured of its custom. Another great advantage of an alliance is that it entails fewer problems of clashes between different cultures, structures and procedures, in management, decision making, remuneration, labour conditions, reporting procedures and norms of conflict resolution, which often turn out to be the biggest obstacles for a successful MA. Of course such clashes can also occur in alliances, but less integration still entails fewer of its problems.

The takeover of a young, dynamic, innovative firm may serve to rejuvenate an old firm (Vermeulen and Barkema 2001). In a growing new firm, the entrepreneur often has to turn him or herself into the role of an administrator, or hire one, to delegate work and institute formal structures and procedures for the coordination of more specialized activities in large-scale production. He or she may not be able or willing to do that, and it may be to the benefit of the firm when it is taken over by a firm with better managerial capability. However, it may be more likely that the entrepreneurial dynamic of the small firm gets stifled in the bureaucracy of the acquirer, in which case it should stay separate.

Table 2.2   Reasons for an MA vs. an alliance

	MA (integration)	Alliance (keeping distance)
Efficiency	Inseparable economy of scale in core activities Inseparable economy of scope	Economy of scale in non-core activities Motivating force of independence Lower costs and risks of integration
Competence	Maintain appropriability, options for future competence Spillover control Rejuvenation Provide management for a growing firm	Maintain focus on core competence Maintain diversity, cognitive distance Maintain entrepreneurial drive
Positional advantage	Control hold-up risk Control quality brand name Protect other partners from spillover Ensure against takeover Keep out competition	Maintain flexibility Maintain local identity/ brand of partner
By default	Partner only available in MA Difficulty of evaluating a take over candidate Collusion forbidden by competition authorities	Partner only available in alliance Interest only in part of a partner MA forbidden by competition authorities
Rule of thumb	*In case of same core competences and same markets*	*In case of complementary competences or markets*

There is an argument of scale for both integration and non-integration. In production, many economies of scale have been reduced, e.g. in computing. However, there is still economy of scale in, for instance, distribution channels, communication networks, network externalities and brand name. For integration, the argument of scale is that one pools volume in activities in which one specializes. For outsourcing, the argument is that for activities that one does not specialize in, an outside, specialized producer can collect more volume, producing for multiple users. That may also offer more opportunities for professional development and career to staff who are specialized in that activity. Note the argument from TCE that if assets are so dedicated that a supplier can produce only for the one user, the scale argument for outsourcing disappears.

There is an argument of scale or scope for integration only if the activities are inseparable (Williamson 1975). It depends on how 'systemic' rather than 'stand-alone' activities are (Langlois and Robertson 1995). There are different forms of economy of scale. One of them, going back to Adam Smith, is greater efficiency by specialization. Often, specialized activities can be outsourced. One form of economy of scope is that different activities share the same underlying fixed cost, for example R&D, management and administration, communication network or brand name. When one of the activities is dropped, the utilization of fixed costs may drop. However, this is not necessarily so. It may be possible to share such overheads with others, as happens, for example, in 'incubators' for small firms, or collaboration in an R&D consortium.

From the perspective of brand image there are arguments for both integration and separation. In an alliance there may be too great a risk that the image or quality of a brand allotted to partners will not be maintained sufficiently scrupulously. On the other hand, it may be better to maintain an independent, outside brand, to preserve its local identity.

Finally, there are default reasons. One would like to adopt one type of organizational form but it is not accessible, because a partner is only available for the other form, or because it is forbidden by competition authorities. In the airline business, for example, MAs are problematic for reasons of national interest, and the fact that landing rights are nationally allocated. Another default is that one would like to take over only part of a larger firm, but it is not separately available for takeover, without the rest, in which one is not interested because it would dilute core competence. Another is that one cannot judge the value of a takeover candidate and needs some period of collaboration in an alliance to find out. Previously, value could more easily be judged by adding up values of material assets than now, when intangibles such as brand name, reputation, skills and knowledge are often more important, and difficult to value.

Thus, the choice between an MA and alliance is quite complex. If one wants a simpler, general rule of thumb, it is as follows: consider full integration, in an MA, only if the partner engages in the same core activities in the same markets. In all other cases, i.e. when activities and/or markets are different, the rule of thumb suggests an alliance. According to this rule, what one would expect, on the whole, is vertical disintegration and horizontal integration.

The theoretical argument is as follows. In horizontal collaboration, with the same activities in the same markets, partners are direct competitors, and it is most difficult to control conflict without integration. The game is more likely to be zero-sum. The temptation to exploit dependence is greatest. There is a threat of direct rather than indirect spillover. In horizontal collaboration core competence is more similar, so that integration does not dilute it too much. Also, here the advantages of alliances are less: the diversity in knowledge is already minimal, with small cognitive distance, and thus there is less need to preserve it by staying apart. Finally, with the same products, technology and markets, differences in culture, structure and procedures are likely to be minimal, and hence problems of integration are less. However, though less than under full integration, such problems can still be substantial.

The argument for this rule is not only theoretical. Bleeke and Ernst (1991) showed empirically that when this rule is applied, the success rate of both MAs and alliances rises substantially. If for a given method of measurement the success rate is less than 50 per cent without the rule, then success rises to 75 per cent with the rule, for both MA and alliances. However, it is emphasized here that the rule given above is only a rule of thumb, to which there are exceptions. For more detailed analysis one can use Table 2.2 and the logic set out previously.

Next to good reasons for MAs, alliances and outsourcing, there are also reasons that are bad, in the sense that they are not in the interests of the firms involved. One such reason is the bandwagon effect: one engages in a practice because it is the fashion to do so. When a practice becomes established, the drive for legitimation may yield pressure to adopt it without much critical evaluation (DiMaggio and Powell 1983). Another reason is a Prisoner's Dilemma that applies especially to an MA: if one does not take over one may be taken over, which may yield a loss of managerial position, so one tries to be the first to take over, even though it would be best for all to stay apart. Another reason is managerial 'hubris': managers want to make a mark and appear decisive or macho. This also applies especially to MAs: those are quicker, more visible and dramatic than the careful buildup of collaboration between independent firms. There is also the often-illusory presumption that a takeover is easier than an alliance. Subsequently, however, the MA often fails due to problems of integration and has to be disentangled again. Even speed is a dubious argument. It may on the surface seem that an MA is in place faster than an alliance, for which one must negotiate longer and set up an elaborate system of 'bilateral governance'. However, the speed of an MA is misleading: the decision may be made quickly, but the subsequent process of integration is often much slower and more problematic than assumed. An alliance is often better even if in the longer run a takeover is the best option, to allow for the process of trust development, discussed before. Also, it yields the option to retract when failure emerges, without too much loss of investment.

An equity joint venture is an intermediate case. It yields advantages of control without full integration of all activities of the parents. Thereby, it allows for more focus on core competences and limits integration problems. It can separate and protect a new, entrepreneurial activity from established bureaucracy. By separating new ventures from the parents one can also better control spillover problems for existing partners. If collaboration entails an activity that remains incorporated in a large, diversified firm, with other activities that may be seen as potentially competitive to a partner, or which maintains connections with the partner's competitors, the spillover risk to them is higher than when the activity is shielded in a joint venture, under partial control by the partner. The new venture may also offer new opportunities for financing.

A well-known question concerning the governance of JVs is whether ownership and control should be symmetric or concentrated in a clear majority shareholder (Killing 1983, Anderson and Gatignon 1986, Geringer and Hebert 1989, Bleeke and Ernst 1991). The argument for the latter is that it yields more decisiveness, which is needed especially under crisis. The argument for the former is that a minority participant may suffer from a 'Calimero syndrome': with less influence and high dependence, he or she may not be motivated to do their best and may be overly suspicious of becoming the victim of opportunism, which blocks the building of trust. A solution may be to separate ownership and management, with symmetric ownership and a rotating majority in decision making, which at any time yields a clear initiative in management for one side, whose actions are monitored by a balanced supervisory board, and disciplined by the perspective that later management will shift to the partner. Another solution is to hire a third party as a manager, under shared control.

Networks

Inter-firm relations go beyond dyads. There may be multiple participants and indirect linkages in networks. Those have implications for the value, risk and governance of relations. One may value a partner not for themselves but for the access that they provide to others (cf. the 'positional advantage' of relations included in Table 2.1). In an alliance, one may need to assess the risk that the partner may be taken over by a competitor, possibly in an indirect way, in which they take over a majority shareholder of the partner (Lorange and Roos 1992). Spillover risk can be indirect, through partners to competitors. If one already has many partners, adding a new one might raise spillover risk for existing partners.

Simmel (1950) showed that a fundamental shift occurs in going from dyadic to triadic relations (see also Krackhardt 1999). In dyadic relations coalitions can occur, and no majority can outvote an individual. In a triad any member by himself has less bargaining power than in a dyad. The threat of exit carries less weight, since the two remaining partners would still have each other. In a triad, conflict is more readily solved. When any two players enter conflict, the third can act as a moderator or 'go-between'.

One stream of literature on networks suggests that players who span 'structural holes' can gain advantage (Burt 1992). If individuals or communities A and B are connected only by C, then C can take advantage of his bridging position by accessing resources that others cannot access, and by playing off A and B against each other. As a result, the third party is maximally powerful and minimally constrained in his actions. This yields Burt's (1992) notion of *tertius gaudens*. Krackhardt pointed out that this principle goes back to Simmel (1950). However, Krackhardt shows that Simmel also indicated that under some conditions the third party is maximally instead of minimally constrained. This occurs when he bridges two different cliques, with dense and strong internal ties, who entertain different values and norms, while both can observe his actions. The third party then has to satisfy the rules or norms of both cliques (the intersection of norm sets), and thereby he is constrained in his actions. The key factor that determines whether the bridging party is minimally or maximally constrained is the degree to which his actions are public, or at least known to both A and B. If not, then the situation described by Burt obtains, and he is minimally constrained. If his actions are public, he is maximally constrained. Membership of multiple cliques then yields a position of potential power, but also constrains the use of it. This often applies to managers or boundary spanners who bridge different departments or organizations. It can apply to the manager of a joint venture who still has allegiance to the parent who remains his employer.

One characteristic of network structure is density: the extent to which participants are linked to each other. With n participants, the maximum number of direct connections, where everyone is connected with everyone else, is $n(n-1)/2$. Thus complexity rises with the square of the number of participants. High density and strong ties yield advantages for governance of relational risk (Coleman 1988). Strong ties yield opportunities for monitoring compliance to agreements, and form the basis for the building of trust. They may be needed for the absorption of knowledge (Hansen 1999). As noted above, one may require relation-specific investments in mutual absorptive capacity, to cross cognitive distance, and this requires a relation of sufficient duration. Dense ties enable an efficient reputation mechanism and opportunities for coalition formation to constrain defectors. Density may be needed to hedge bets concerning the future presence of sources of information, in volatile networks of innovation, to use third parties to supplement one's absorptive capacity and to test the reliability of information (in 'triangulation') (Nooteboom and Gilsing 2004).

On the other hand, with high density there is much redundancy of contacts, which can reduce efficiency, raise confusion, yield duplication of knowledge and effort and waste resources in excessive coordination (Burt 1992; Granovetter 1973). High density yields access to varied sources of knowledge, but the downside of it may be high risk of spillover. High density of strong ties also reduces flexibility of configuration: there are more connections one has to get out of, and coalitions may prevent exit. Strong, durable ties, with frequent interaction, yield a reduction of cognitive distance, and thus constrain the variety needed for learning. Finally, from complexity theory we learn that dense structure can yield reverberations, with negative or positive feedbacks in the network, which can yield unforeseeable, chaotic outcomes.

Hierarchies serve to avoid problems of density. An alternative is a hub-and-spoke system, with the central hub collecting and distributing knowledge from all to all. This entails the notion of centrality: the extent to which there are participants with a large number of direct connections compared to others. Centrality in a network increases one's power of bridging structural holes, offering partners opportunities for indirect access to competence, knowledge, or markets. However, this can also raise spillover risks for partners. Central power limits flexibility of reconfiguring network relations. It also subjects the central agent to more norm sets that may be in conflict, which may constrain his actions (see the Simmelian argument discussed before).

Propositions

> **H11** There are many contingencies to be taken into account in the choice between MA and alliance. There are good reasons for both. A simple rule of thumb, with its exceptions, is to engage in MA only when partners have the same products in the same market. In all other cases the base case is an alliance.
>
> **H12** One should look beyond dyads to network structure and position, to take into account redundancy, indirect spillover, risks of a takeover of partners, reputation effects, limits to flexibility and too much reduction of cognitive distance.
>
> **H13** Third parties bridging structural holes can be both minimally and maximally constrained, depending on whether their conduct is public or not.

GOVERNANCE

Instruments

Instruments for governance, i.e. for controlling relational risk, derive from several theoretical perspectives. TCE yields the instruments of mutual dependence, sharing ownership and risk of specific assets, and 'hostages'. Social exchange theory yields insights in the basis, limits and process of trust. Social network theory yields insights in the role of network structure and position, and third parties. The instruments are summarized in Table 2.3. Each instrument has its drawbacks or limitations.

The first two instruments in the table fall outside the scope of alliances. The first entails a cop-out: in view of relational risks, hold-up is avoided by not engaging in dedicated investments, and spillover is avoided by not giving away any sensitive knowledge. The opportunity cost of this is that one may miss opportunities to achieve high added value in the production of specialties by investing in collaboration and learning with partners. The second instrument is integration in an MA, with the advantages and drawbacks discussed above. In the second part of Table 2.3, we find the

Table 2.3   Instruments of governance and their drawbacks

Instrument	Drawback
Risk avoidance: no specific investments, no knowledge transfer	Lower added value, with lesser product differentiation (in case of dedicated technology); No learning
Integration: MA	Less flexibity, variety, motivation, problems of integration (see Table 2.2)
Number of partners: Maintain alternatives Demand exclusiveness	Mutiple set-up costs, spillover risk for partners Limitation of variety learning
Contracts:	Problematic under uncertainty, can be expensive, Straight-jacket in innovation, can generate distrust
Self-interest: Mutual dependence hostages, reputation	Opportunistic: requires monitoring and is sensitive to change of capabilities, conditions, entry of new players
Trust:	Needs building up if not already present has limits, how reliable?
Go-betweens:	May not be available, how reliable?
Network position:	Needs time to build, side-effects

instruments for alliances, where one accepts the problems of dependence due to dedicated investment and possibilities of spillover, and seeks to control them by means other than full integration.

One instrument for the control of hold-up risk is to maintain multiple partners, in order not to become dependent on any one of them. However, maintaining relations with alternative partners entails a multiplication of costs in dedicated investments and the governance needed to control the risks involved. By having those relations one increases the spillover risk for partners. As a result, none of them may be willing to give sensitive information, which degrades their value as sources of complementary competence and learning.

To prevent spillover, one may insist on exclusiveness from any partner, in the specific area of collaboration involved, i.e. demand that they do not associate with one's potential competitors in that area. The first problem with this is that the demand of exclusiveness blocks the variety of the partner's sources of learning, which reduces their value as a partner in learning, at a cognitive distance that is maintained by their interaction with outside contacts. Hence one should consider whether spillover is really a significant risk, as discussed before. If it is not, all parties can gain from maintaining multiple partners, not so much for maintaining bargaining position as for maintaining variety of sources of learning and flexibility of configurations.

Another instrument is a contract, in an attempt to close off 'opportunities for opportunism', by contracts. The problem with this instrument is fourfold. It can be expensive to set up. It can be ineffective for lack of possibilities to monitor compliance, due to asymmetric information. It can be impossible because of uncertainty concerning future contingencies that affect contract execution. This applies especially when the purpose of collaboration is innovation. Finally, detailed contracts for the purpose of closing off opportunities for opportunism express distrust, which can raise reciprocal suspicion and distrust, with the risk of ending up in a vicious circle of regulation and distrust that limits the scope for exploration of novelty and obstructs the build-up of trust as an alternative approach to governance.

Another approach is to aim at the self-interest of the partner and limit incentives to utilize any opportunities for opportunism left by incomplete contracts. These instruments have been mostly developed in TCE. Self-interest may arise from mutual dependence, in several ways. One is that the partner participates more or less equally in the ownership and hence the risk of dedicated assets. A second is that one uses one's own dedicated investments to build and offer a unique, valuable competence to the partner. Thus, the effect of dedicated investments can go in different directions: it makes one dependent due to switching costs, but it can also make the partner dependent by offering them high and unique value. This instrument can yield an upward spiral of value, where partners engage in a competition to be of unique value for each other. Dependence also arises from a 'hostage', as also suggested by TCE. One form of 'hostage' is minority participation, where one can sell one's shares to someone who is eager to take over the partner. A more prevalent form is sensitive information. Here, the notion of 'hostage' connects with the notion of spillover. One may threaten to pass on sensitive knowledge to a partner's competitor. Reputation also is a matter of self-interest: one behaves well in order not to sacrifice potentially profitable relations with others in the future (Weigelt and Camerer 1988).

The limitation of instruments aimed at self-interest is that they are not based on intrinsic motivation, and require monitoring, which may be difficult, especially in innovation. Furthermore, balance of mutual dependence is sensitive to technological change and to the entry of new players that might offer more attractive partnerships. 'Hostages' may not be returned in spite of compliance to the agreement. Reputation mechanisms may not be in place, or may work imperfectly (Hill 1990; Lazaric and Lorenz 1998). They require that a defector cannot escape or dodge a breakdown of reputation, perhaps because they are selling the business or switching to another industry or another country. It requires that complaints of bad behaviour be checked for their truth and communicated to potential future partners of the culprit.

Trust

Trust is a slippery and complex notion, which cannot be dealt with completely in this chapter. For a more detailed, systematic account, see Nooteboom (2002), from which a number of key features are summarized here. A common definition of trust is that it entails an acceptance of relational risk, in the expectation that no harm will be done. Most people would agree that one couldn't speak of 'real' trust when the expectation is based on coercion by hierarchy, contract or incentives of self-interest. Real trust is then defined as the expectation that no harm will be done, even though the partner has both the opportunity and the incentive to defect (Bradach and Eccles 1989; Chiles and McMackin 1996; Nooteboom 1999). Distrust means that one does not expect a partner to conform to expectations or agreements unless they are forced to do so or has a strong interest in doing so.

Trust can be based on 'universalistic' ethics, values and norms of reciprocity or obligation that prevail in a culture, on characteristics such as membership of a family or association, in 'characteristics-based trust' (Zucker 1986), or on 'particularistic' bonding in a specific relationship, on the basis of empathy or mutual identification, friendship or habitualization (Gulati 1995; Lewicki and Bunker 1996; McAllister 1995; Nooteboom et al. 1997).

Another important point is that trustworthiness will almost always have its limits. Williamson (1993) was surely correct to say that blind, unconditional trust is generally unwise. Even the sincerest wish to be loyal may break down under temptation or pressures of survival. One cannot really expect that even one's best friend will remain loyal under the direst pressures of survival, such as torture, and it would be unethical to demand that. Thus, trust is subject to tolerance levels: one trusts within limits of observed conditions and actions. Within those limits, one does not continually question competences and loyalties, and one is not continually on the lookout for opportunities for opportunism, for the partner and for oneself. When the limits are exceeded, one starts to be aware of potential opportunism, and to consider exit. In voice,[1] such concerns are communicated, in an effort to jointly resolve them. Nevertheless, voice is constrained by the possibility of exit, as Hirschman recognized. However, though this may seem paradoxical, one can exit also with the use of voice. This point will return later.

If trust is not in place prior to collaboration, it has to be built up in the relation, in 'process-based trust'. One cannot buy and install trust, but one can create the conditions for it to develop, in 'trust-sensitive management' (Sydow 2000, p. 54). As a relationship develops, partners begin to know each other better, and can better assess the extent and limits of trustworthiness ('knowledge-based trust'). Convergence of cognitive frameworks may arise, which can lead to mutual identification ('identification-based trust') (Lewicki and Bunker 1996; McAllister 1995). Partners understand and can identify with each other's goals, weaknesses and mistakes, and are able to engage in the give and take of voice. This does not entail that they always agree. There may be sharp disagreements, but those are combined with a willingness to express and discuss them more or less openly, in 'voice', offering mutual benefit of the doubt. As a result, when conflicts are jointly resolved they will deepen trust rather than breaking it.

Lewicki and Bunker claimed that when trust is not in place at the beginning, one has to start with control, and later switch to trust based on the growth of empathy and identification. However, as noted before, control in the form of detailed contracts to close off opportunities for opportunism can yield an escalation of mutual distrust that precludes the building of trust. Alternatively, one could start a relationship with small steps of low-level mutual dependence, and increase commitment as trust in competence and commitment grows (Shapiro 1987). One limitation of this is that one may not have the time to build trust in this way. The process of trust-building may be speeded up by the use of intermediaries. This is discussed in a later section.

Mutual openness is needed to pool complementary assets and competences. It is also essential to the building of trust (Zand 1972). A rich flow of information is needed for the 'let's work things out' approach of the voice strategy (Maguire et al. 2001). Things may go wrong in a relationship because of outside accidents, mistakes, lack of competence or because of opportunism, but in practice they are difficult to identify because an opportunist will claim mishaps or mistakes as the cause of disappointing results. That is why openness also about one's mistakes is crucial, to prevent their being interpreted as opportunism.

Strategies of Outsourcing

A traditional strategy of purchasing is to go for minimal cost, in competition between suppliers, and maximum flexibility of switching between suppliers. This tends to lead to lack of openness on the part of suppliers: to maintain bargaining power on price, a supplier will not surrender information on competences (and lack of them), cost structure, capacity utilization and order portfolio. This prevents close cooperation in joint development and production. Without guarantee of continued business a supplier will not engage in relation-specific investments. This strategy prevents the achievement of high added value, in specialty products that benefit from the utilization of opportunities for pooling knowledge and competence. That is fine for standardized commodities that lack scope for specialties and innovation.

However, for specialty products, innovation and utilization of complementary dynamic capabilities, in buyer–supplier relationships, one needs to achieve the openness that is needed in co-makership and early supplier involvement, to build and maintain mutual understanding and trust. Then, it is difficult to achieve openness when the focus is on bargaining for price, to secure sufficient profit. Therefore, the buyer must grant his suppliers profit, in 'price-minus costing': production cost is set at price minus a profit margin for the supplier. One argument for this is that there is lack of continuity in a partner who may go broke for lack of profit. Another is to 'earn' the partner's openness. What may be lost in price may be more than recovered in quality and speed. This is particularly so when contracts cannot be closed, and a partner regains profit in unforeseen, uncontracted work. An example of this is the building industry. Bargaining focuses on minimum price, which yields closure of information, which inhibits opportunities to collaborate for optimal quality, speed, and fit in time, place and technical interfaces. Insiders claim that up to 25 per cent of final cost is due to such mismatches. And then the builder recoups profit in unforeseen work, so that minimum cost is not actually achieved.

Open-book contracting and price-minus costing is one of the lessons the West has learned from the Japanese, particularly in the car industry (Cusumano and Takeishi 1991; Dyer 1996; Dyer and Ouchi 1993; Helper 1990; de Jong and Nooteboom 2000; Kamath and Liker 1994; Lamming 1993).

However, Western buyers did not copy the all-too-durable and exclusive buyer–supplier relations that were customary in the Japanese vertical structures of *keiretsu*. As discussed before, some durability is needed to recoup dedicated investments, and exclusiveness may be needed to control spillover, but too durable and exclusive relations yield rigidities. The West adapted the system, with relations that are closer to optimal duration, and without unnecessary exclusiveness, in view of limited risks of spillover. Now the Japanese appear to be learning from that and are beginning to open up or break down their *keiretsu* (de Jong and Nooteboom 2000).

Go-betweens

One can use the services of third parties, 'guardians of trust' (Shapiro 1987), 'intermediate communities' (Fukuyama 1995), or 'go-betweens' (Nooteboom 1999). One role, recognized in TCE, is that of arbitration or mediation in 'trilateral governance'.

A second role is to assess the value of information before it is traded, to solve the 'revelation problem': if one wants to sell information, the partner will want to assess its value by looking into it, but then he already has the information and might no longer pay for it. This problem can be solved in several ways. One is the use of licences, with a limited payment up front, and later payment in proportion to the emerging yields of the information. Another is to let a go-between assess the value of the information for the potential buyer.

A third role for a go-between is to create mutual understanding, helping to cross cognitive distance. A fourth role is to monitor information flow as a guard against spillover. The reason for this role is that if partner A does the monitoring himself, in the firm of partner B, then this look inside B's firm may increase the spillover from B to A. A fifth role is to act as a guardian of 'hostages'. Without that, there may be a danger that the 'hostage' keeper does not return the 'hostage' even if the partner sticks to the agreement. The third party has an interest in maintaining symmetric trust and acceptance by both protagonists.

A sixth, and perhaps most crucial, role is to act as an intermediary in the building of trust. Trust relations are often entered with partners who are trusted partners of someone you trust (Sydow 2000). If X trusts Y and Y trusts Z, then X may rationally give trust in Z a chance. X needs to feel that Y is able to judge well and has no intention to lie about his judgement. This can speed up the building of trust between strangers, which might otherwise take too long. This is particularly important in view of the dynamics of the build-up and breakdown of trust. It was noted above that new relationships might have to start small, with low stakes that are raised as trust builds up. This may be needed especially when contracts are not feasible or desirable, as is the case particularly in innovation. As indicated, the disadvantage of such a procedure is its slowness. A go-between may provide help for a more speedy development. Intermediation in the first small and ginger steps of cooperation, to ensure that they are successful, can be very important in the building of a trust relation. The intermediary can perform valuable services in protecting trust when it is still fragile: to eliminate misunderstanding and allay suspicions when errors or mishaps are mistaken for signals of opportunism.

A seventh role, related to the sixth, is to help in the timely and least destructive disentanglement of relations. To eliminate misunderstanding, to prevent acrimonious and mutually damaging battles of divorce, a go-between can offer valuable services, to help in 'a voice type of exit'. This latter concept is discussed later. An eighth role is to support a reputation mechanism. For a reputation mechanism to work, infringement of agreements must be observable, its report must be credible, and it must reach potential future partners of the culprit. The go-between can help in all respects.

Contingencies

There is no single and universal best recipe for managing inter-organizational relationships (IOR). One will generally select some combination of mutually compatible and supporting instruments from the toolbox of governance, and the use of a single instrument will be rare. The choice and effectiveness of instruments depend on conditions: the goals of collaboration, characteristics of the participants, technology, markets and the institutional environment. For example, there is no sense in contracts when the appropriate institutions of laws are not in place, the police or judiciary is corrupt, and when compliance cannot be monitored. When technology is flexible, so that one can produce a range of different specific products with one set-up, the specificity of investments and hence the problem of hold-up is limited. Possibilities of spillover are constrained when knowledge is tacit, and do not matter when technology changes fast. Reputation mechanisms don't work when there are ample exit opportunities for defectors. Trust is difficult in a distrustful environment, where cheating rather than loyalty is the norm.

Innovation has its special conditions. Exchange of knowledge is crucial, with corresponding risks of spillover. In innovation especially, the competencies and intentions of strangers are difficult to judge. Relevant reputation has not yet been built up. Uncertainty is large, limiting the possibility of specifying the contingencies of a contract. Specific investments are needed to set up mutual understanding. There is significant hold-up risk. Detailed contracts would limit the variety and scope of unpredictable actions and initiatives that innovation requires. With these problems in contracts, trust is most needed to limit relational risk. An additional problem with contracts is that they may obstruct the building of trust. This does not mean that there are no contracts, but that they cannot be too detailed with the purpose of controlling hold-up risk. A productive combination of instruments is mutual dependence complemented by trust on the basis of an emerging experience in competent and loyal collaboration. Trust is needed besides mutual dependence, because the latter is sensitive to changing conditions. Trust is more difficult under asymmetric dependence because the more dependent side may be overly suspicious (Klein Woolthuis 1999), in the 'Calimero syndrome' mentioned before. In all this, go-betweens can help, with the roles specified before. Without them the building of trust may be too slow.

In the literature, contracts and trust are primarily seen as substitutes. Less trust requires more contracts, and detailed contracts can obstruct the building of trust. However, this view is too simplistic. Trust and control can also be complements (Das and Teng 1998; 2001; Klein Woolthuis 1999; Klein Woolthuis et al. 2005). There may be a need for an extensive contract, not so much to foreclose opportunities for opportunism, but to serve as a record of agreements in a situation where coordination is technically complex. A simple contract may provide the basis for building trust. One may need to build up trust before engaging in the costs and risks of setting up a contract. Signing a contract may be a ritual of agreement and a symbol of trust.

Finally, perhaps the most important point is that relationships should be seen as processes rather than entities that are instituted and left to themselves. Conditions may change. A frequent problem is that a relationship starts with a balance of dependence, but in time the attractiveness of one of the partners slips, due to slower learning, appropriation of their knowledge by the other, or institutional, technological or commercial change. A classic example is the alliance between an American company that supplies design or technology, and a Japanese company that supplies access to the Japanese market. After a while, the Japanese company has copied the technology and competences, but is itself still needed for market access. This often leads to a disentangling of the alliance or one side buying out the other. Alliances may from the start have been designed for a limited duration. An alliance may be used to have a joint standard accepted in the market, or as an exploratory stage for an MA, to assess value and the viability of integrating cultures.

One needs to select instruments of governance with a view to needs and opportunities for adapting or replacing them in the future. One example is not to go for detailed contracts to foreclose opportunities for opportunism, in order to maintain an opening for the building of trust. An aggressive stance of exit is often difficult to credibly turn around in a collaborative stance of voice. In the process of setting up collaboration a mistake often is that it starts with harsh negotiation between the leaders, and then is thrown into the lap of an implementation manager, who is saddled with an atmosphere of distrust and acrimony that jeopardizes the relationship before it has properly started. A solution is to include the implementation manager in the negotiation process, to charge negotiators with implementation as well, or to make use of a skilled go-between. If earlier in the relation one took a voice approach, then an adversarial approach to ending the relationship is likely to create more upheaval than if one took an exit stance to the relationship from the start.

Ending Relationships

If it is normal and desirable that relationships should end, before they create rigidities, the question arises how to do that. That may be at least as important as beginning a relationship, and is probably more difficult. If one wants to end a relationship because a more attractive option has emerged, should one announce this intention at an early stage, or should one prepare one's exit in secret, and drop the bomb on the partner at the last moment? In other words should one go for an adversarial or a collaborative 'voice' mode of exit (Nooteboom 1999)? With the collaborative mode, one offers the partner a timely way out, with minimal damage so that they stop making specific investments that would potentially increase their switching cost, and one helps them to find a new partner, in order to minimize disruption. A harsh exit approach to end a relationship may yield considerable problems of emotional response, acrimony, resistance, and damage to reputation. A voice approach to exit does give the partner time to better obstruct one's departure. Nevertheless, collaborative divorce is viable if the partner can be expected to cut their losses and welcome the help to get out with minimal damage.

Propositions

H14 There is no universal best tool for governance. There is a toolbox from which one should judiciously select a mix that fits with a range of contingencies.

H15 In innovation in particular, formal control or deterrence has its limits, and mutual dependence combined with trust is to be preferred.

H16 Trust is viable, but is subject to limits.

H17 Trust and control are both complements and substitutes.

H18 Go-betweens can help to solve a range of problems in governance.

H19 Voice is constrained by exit, but there are voice forms of ending a relationship.

H20 Relationships are dynamic, and instruments of governance should be also chosen with a view to later adjustment.

CONCLUSION

This chapter offers an integrated set of theoretical perspectives, goals, forms of collaboration and instruments for its governance. They form a toolbox from which one can select according to circumstances. Competence and governance views are integrated. Transaction cost economics suffers from fundamental shortcomings, but still retains elements that are useful for governance. Outsourcing everything that does not belong to core competence can be problematic. One needs to consider loss of options on future core competences and relational risk. However, some of these problems can also be solved in alliances. Voice is limited by exit, but there are voice forms of ending a relationship. Ending relationships well is as important as setting them up.

The analysis yields tools for practice: for deciding what to outsource, whether to go for a merger/acquisition or alliance, what instruments to use for governance, and under what conditions. A selection of policy implications is given below.

- Diversity and cognitive distance carry costs and risks but also opportunities for learning.
- Relations should be stable enough to encourage specific investments, e.g. in mutual understanding and trust, but not so long as to yield rigidities that block innovation.
- There are too many mergers and acquisitions, and they should, in principle, be limited to situations of similar products in the same markets.

- Under the uncertainties of innovations in particular, detailed contracts are problematic, and there is more need for trust.
- Trust can go beyond calculative self-interest, but has its limits, depending on external temptations and pressures of survival.
- For both opportunities and risks one should look beyond direct relationships to indirect ties in networks.
- In networks, dense and strong ties may be needed for innovation, but they can also constrain it.

Much of the logic set out here can also be applied to relations within firms, but that is not the focus of this chapter.

There is much more research to be done on the process side of trust and collaboration. In further research, much might be gained by the use of social psychology. This could yield more insight in how people perceive and interpret actions of others, and infer or attribute motives and capabilities to partners (Nooteboom 2002). In networks, more research is needed to find out, in more detail, when density and strength of ties are good and when they are bad for innovation. For this, we need to consider different dimensions of tie strength.

NOTE

1. Hirschman (1970) makes a distinction between alternative ways of reacting to dissatisfaction with organizations. One, exit, is for the member to quit the organization or for the customer to switch to the competing product, and the other, voice, is for members or customers to agitate and exert influence for change 'from within'.

REFERENCES

Anderson, E. and H. Gatignon (1986), 'Modes of foreign entry: a transaction cost analysis of propositions', *Journal of International Business Studies*, **17** (3), 1–26.

Beije, P. (1998), 'Technological cooperation between customer and subcontractors', ERASM research report 98-39, Rotterdam School of Management, Erasmus University Rotterdam.

Berger, J., N.G. Noorderhaven and B. Nooteboom (1995), 'The determinants of supplier dependence: an empirical study', in J. Groenewegen, C. Pitelis and S.E. Sjöstr (eds), *On Economic Institutions: Theory and Applications*, Aldershot, UK and Brookfield, USA: Edward Elgar, pp. 195–212.

Bettis, R., S. Bradley and G. Hamel (1992), 'Outsourcing and industrial decline', *Academy of Management Executive*, **6** (1), 7–16.

Blackler, F. (1995), 'Knowledge, knowledge-work and organizations: an overview and interpretation', *Organization Studies*, **16** (6), 1021–46.

Bleeke, J. and D. Ernst (1991), 'The way to win in cross-border alliances', *Harvard Business Review*, **69** (6), 127–35.

Bradach, J.L. and R.G. Eccles (1989), 'Plural authority, and trust: from ideal types to plural forms', *Annual Review of Sociology*, **15** (August), 97–118.

Bromiley, P. and L.L. Cummings (1992), 'Transaction costs in organizations with trust', working paper, Carlson School of Management: University of Minnesota, Minneapolis.

Burt, R.S. (1992), *Structural Holes: The Social Structure of Competition*, Cambridge, MA: Harvard University Press.

Carter, R. (2002), 'Empirical work in transaction cost economics: critical assessments and alternative interpretations', PhD dissertation, Cambridge University.

Chiles, T.H. and J.F. McMackin (1996), 'Integrating variable risk preferences: trust and transaction cost economics', *Academy of Management Review*, **21** (1), 73–99.

Choo, C.W. (1998), *The Knowing Organization*, Oxford: Oxford University Press.

Cohen, W. and D.A. Levinthal (1990), 'Absorptive capacity: a new perspective on learning and innovation', *Administrative Science Quarterly*, **35** (1), 128–52.

Coleman J.S. (1988), 'Social capital in the creation of human capital', *American Journal of Sociology*, **94** (supplement), 95–120.

Colombo, M.G. and P. Garrone (1998), 'Common carriers' entry into multimedia services', *Information Economics and Policy*, **10** (1), 77–105.

Contractor, F.J. and P. Lorange (1988), 'Why should firms cooperate? The strategy and economic basis for cooperative ventures', in F.J. Contractor and P. Lorange (eds), *Cooperative Strategies in International Business*, Lexington, MA: Lexington Books, pp. 3–30.

Cusumano, M.A. and A. Takeishi (1991), 'Supplier relations and management: a survey of Japanese, Japanese-transplants, and US auto plants', *Strategic Management Journal*, **12** (7), 563–88.

Das, T.K. and B.S Teng (1998), 'Between trust and control: developing confidence in partner cooperation in alliances', *Academy of Management Review*, **23** (3), 491–512.

Das, T.K., and B.S. Teng (2001), 'Trust, control and risk in strategic alliances: an integrated framework', *Organization Studies*, **22** (2), 251–84.

De Jong, G. and B. Nooteboom (2000), *The Causal Structure of Long-term Supply Relationships*, Deventer, The Netherlands: Kluwer.

Deutsch, M. (1973), *The Resolution of Conflict: Constructive and Destructive Processes*, New Haven, CT: Yale University Press.

DiMaggio, P.J. and W.W. Powell (1983), 'The iron cage revisited: institutional isomorphism and collective rationality in organizational fields', *American Sociological Review*, **48** (2), 147–160.

Dyer, J.H. (1996), 'Specialized supplier networks as a source of competitive advantage: evidence from the auto industry', *Strategic Management Journal*, **17** (4), 271–91.

Dyer, J.H. and W.G. Ouchi (1993), 'Japanese-style partnerships: giving companies a competitive edge', *Sloan Management Review*, **35** (1), 51–63.

Faulkner, D. (1995), *International Strategic Alliances: Cooperating to Compete*, Maidenhead: McGraw-Hill.

Fukuyama, F. (1995), *Trust: The Social Virtues and the Creation of Prosperity*, New York: Free Press.

Gambetta, D. (1988), 'Can we trust in trust?', in D. Gambetta (ed.), *Trust: Making and Breaking Cooperative Relations*, Oxford: Blackwell, pp. 213–37.

Gargiulo, M. and M. Benassi (1999), 'The dark side of social capital', in R.Th. A. J. Leenders and S.M. Gabbay (eds), *Corporate Social Capital and Liability*, Dordrecht: Kluwer, pp. 298–322.

Geringer, M.J. and L. Hebert (1989), 'Control and performance of international joint ventures', *Journal of International Business Studies*, **20** (2), 235–54.

Granovetter, M.S. (1973), 'The strength of weak ties', *American Journal of Sociology*, **78** (6), 1360–80.

Granovetter, M.S. (1985), 'Economic action and social structure: a theory of embeddedness', *American Journal of Sociology*, **91** (3), 481–510.

Granstrand, O., P. Patel and K. Pavitt (1997), 'Multi-technology corporations: why they have "distributed" rather than "distinctive core" competencies', *California Management Review*, **39** (4), 8–25.

Grey, C. and C. Garsten (2001), 'Trust, control and the post-bureaucracy', *Organization Studies*, **22** (2), 229–50.

Gulati, R. (1995), 'Does familiarity breed trust? The implications of repeated ties for contractual choice in alliances', *Academy of Management Journal*, **30** (1), 85–112.

Hansen, M.T. (1999), 'The search–transfer problem: the role of weak ties in sharing knowledge across organization subunits', *Administrative Science Quarterly*, **44** (1), 82–111.

Helper, S. (1987), 'Supplier relations and innovation: theory and application to the US auto industry', PhD thesis, Harvard University.

Helper, S. (1990), 'Comparative supplier relations in the US and Japanese auto industries: an exit/voice approach', *Business and Economic History*, **19**, 1–10.

Hennart, J.F. (1988), 'A transaction cost theory of equity joint ventures', *Strategic Management Journal*, **9** (4), 361–74.

Hill, C.W.L. (1990), 'Cooperation, opportunism and the invisible hand: implications for transaction cost theory', *Academy of Management Review*, **15** (3), 500–13.

Hirschman, A.O. (1970), *Exit, Voice and Loyalty: Responses to Decline in Firms, Organizations and States*, Cambridge, MA: Harvard University Press.

Jarillo, J.C. (1988), 'On strategic networks', *Strategic Management Journal*, **9** (1) 31–41.

Kamath, R.R. and J.K. Liker (1994), 'A second look at Japanese product development', *Harvard Business Review*, **72** (6), 154–70.

Killing, J.P. (1983), *Strategies for Joint Ventures*, New York: Praeger.

Klein Woolthuis, R. (1999), 'Sleeping with the enemy: trust, dependence and contracts in inter-organizational relationships', doctoral dissertation, Twente University, Enschede, The Netherlands.

Klein Woolthuis, R., B. Hillebrand and B. Nooteboom (forthcoming), 'Trust, contract and relationship development', *Organization Studies*.

Kolb, D. (1984), *Experiential Learning: Experience as the Source of Learning and Development*, Englewood Cliffs, NJ: Prentice Hall.

Krackhardt, D. (1999), 'The ties that torture: Simmelian tie analysis in organizations', *Research in the Sociology of Organizations*, **16**, 183–210.

Lamming, R. (1993), *Beyond Partnership*, New York: Prentice Hall.

Langlois, R.N. and P.L. Robertson (1995), *Firms, Markets and Economic Change: A Dynamic Theory of Business Institutions*, London: Routledge.

Lazaric, N. and E. Lorenz (1998), 'The learning dynamics of trust, reputation and confidence', in N. Lazaric and E. Lorenz (eds), *Trust and Economic Learning*, Cheltenham, UK and Lyme, USA: Edward Elgar, pp. 1–22.

Leenders, R.Th. A.J. and S.M. Gabbay (1999), 'Corporate social capital: the structure of advantage and disadvantage', in R.Th. A.J. Leenders and S.M. Gabbay (eds), *Corporate Social Capital and Liability*, Dordrecht: Kluwer, pp. 1–14.

Lewicki, R.J. and B.B. Bunker (1996), 'Developing and maintaining trust in work relationships', in R.M. Kramer and T.R. Tyler (eds), *Trust in Organizations: Frontiers of Theory and Research*, Thousand Oaks, CA: Sage Publications, pp. 114–39.

Lippman, S. and R.P. Rumelt (1982), 'Uncertain imitability: an analysis of inter-firm differences in efficiency under competition', *Bell Journal of Economics*, **13** (2), 418–38.

Lorange, P. and J. Roos (1992), *Strategic Alliances*, Cambridge: Blackwell.

Macaulay, S. (1963), 'Non-contractual relations in business: a preliminary study', *American Sociological Review*, **28** (1), 55–67.

Macneil, I. (1980), *The New Social Contract: An Enquiry into Modern Contractual Relations*, London: Yale University Press.

Maguire, S., N. Philips and C. Hardy (2001), 'When "Silence = death", keep talking: trust, control and the discursive construction of identity in the Canadian HIV/AIDS treatment domain', *Organization Studies*, **22** (2), 285–310.

McAllister, D.J. (1995), 'Affect- and cognition-based trust as foundations for inter-personal cooperation in organizations', *Academy of Management Journal*, **38** (1), 24–59.

Mol, M. (2001), 'Outsourcing, supplier relations and internationalization: global sourcing strategy as a Chinese puzzle', ERIM PhD Series in Management 10, Erasmus University Rotterdam.

Murakami, Y. and T.P. Rohlen (1992), 'Social-exchange aspects of the Japanese political economy: culture, efficiency and change', in S. Kumon and H. Rosorsky (eds), *The Political Economy of Japan*, vol. 3, *Cultural and Social Dynamics*, Stanford, CA: Stanford University Press, pp. 63–105.

Noorderhaven, N.G. (1996), 'Opportunism and trust in transaction cost economics', in J. Groenewegen (ed.), *Transaction Cost Economics and Beyond*, Boston, MA: Kluwer, pp. 105–28.

Nooteboom, B. (1992), 'Towards a dynamic theory of transactions', *Journal of Evolutionary Economics*, **2** (4), 281–99.

Nooteboom, B. (1999), *Inter-firm Alliances: Analysis and Design*, London: Routledge.

Nooteboom, B. (2000), *Learning and Innovation in Organizations and Economies*, Oxford: Oxford University Press.

Nooteboom, B. (2002), *Trust: Forms, Foundations, Functions, Failures, and Figures*, Cheltenham, UK and Northampton, MA, USA: Edward Elgar.

Nooteboom, B. (2004a), 'Governance and competence: how can they be combined?', *Cambridge Journal of Economics*, **28** (4), 505–26.

Nooteboom, B. (2004b), *Inter-firm Collaboration, Learning and Networks: An Integrated Approach*, London: Routledge.

Nooteboom, B. and V.A. Gilsing (2004), 'Density and the strength of ties in innovation networks', Eindhoven Centre for Innovation Studies working paper 04.01, Eindhoven University of Technology.

Nooteboom, B., J. Berger and N.G. Noorderhaven (1997), 'Effects of trust and governance on relational risk', *Academy of Management Journal*, **40** (2), 308–38.

Ohmae, K. (1989), 'The global logic of strategic alliances', *Harvard Business Review*, **67** (2), 143–54.

Ouchi, W.G. (1980), 'Markets, bureaucracies, and clans', *Administrative Science Quarterly*, **25** (1), 129–41.

Parkhe, A. (1993), 'Strategic alliance structuring: a game theoretic and transaction cost examination of inter-firm cooperation', *Academy of Management Journal*, **36** (4), 794–829.

Penrose, E. (1959), *The Theory of the Growth of the Firm*, New York: Wiley.

Porter, M.E. and M.B. Fuller (1986), 'Coalitions and global strategies', in M.E. Porter (ed.), *Competition in Global Industries*, Boston, MA: Harvard Business School Press, pp. 315–44.

Powell, W.W. (1990), 'Neither market nor hierarchy: network forms of organization', in B.M. Staw and L.L. Cummings (eds), *Research in Organizational Behavior*, vol 12, Greenwich, CT: JAI Press, pp. 295–336.

Ring, P. and A. van de Ven (1992), 'Structuring cooperative relationships between organizations', *Strategic Management Journal*, **13** (2), 483–98.

Ring, P. and A. van de Ven (1994), 'Developmental processes of cooperative inter-organizational relationships', *Academy of Management Review*, **19** (1), 90–118.

Sako, M. (1992), *Prices, Quality, and Trust: Inter-Firm Relations in Britain and Japan*, Cambridge: Cambridge University Press.

Shapiro, S.P. (1987), 'The social control of impersonal trust', *American Journal of Sociology*, **93** (3), 623–58.

Simmel, G. (1950), 'Individual and society', in K.H. Wolff (ed.), *The Sociology of George Simmel*, New York: Free Press.

Smircich, L. (1983), 'Organization as shared meaning', in L.R. Pondy, P.J. Frost, G. Morgan and T.C. Dridge (eds), *Organizational Symbolism*, Greenwich, CT: JAI Press, pp. 55–65.

Stoelhorst, J.W. (1997), 'In search of a dynamic theory of the firm', doctoral dissertation, Twente University, The Netherlands.

Sydow, J. (2000), 'Understanding the constitution of interorganizational trust', in C. Lane and R. Bachmann (eds), *Trust In and Between Organizations: Conceptual Issues and Empirical Applications*, Oxford: Oxford University Press, pp. 31–63.

Teece, D.J. (1986), 'Profiting from technological innovation: implications for integration, collaboration, licensing and public Policy', *Research Policy*, **15** (6), 285–305.

Uzzi, B. (1996), 'The sources and consequences of embeddedness for the economic performance of organizations: the network effect', *American Sociological Review*, **61** (4), 674–98.

Vermeulen, F. and H. Barkema (2001), 'Learning through acquisitions', *Academy of Management Journal*, **44** (3), 457–76.

Weick, K.F. (1979), *The Social Psychology of Organizing*, Reading, MA: Addison-Wesley.

Weick, K.F. (1995), *Sensemaking in Organizations*, Thousand Oaks, CA: Sage.

Weigelt, K, and C. Camerer (1988), 'Reputation and corporate strategy: a review of recent theory and applications', *Strategic Management Journal*, **9** (5), 443–54.

Williamson, O.E. (1975), *Markets and Hierarchies*, New York: Free Press.

Williamson, O.E. (1985), *The Economic Institutions of Capitalism: Firms, Markets, Relational Contracting*, New York: Free Press.

Williamson, O.E. (1993), 'Calculativeness, trust, and economic organization', *Journal of Law and Economics*, **36** (first of two parts), 453–86.

Williamson, O.E. (1999), 'Strategy research: governance and competence perspectives', *Strategic Management Journal*, **20** (12), 1087–108.

Wuyts, S., M. Colombo, S. Dutta and B. Nooteboom (forthcoming), 'Empirical tests of optimal cognitive distance', *Journal of Economic Behavior and Organization*.

Zand, D.E. (1972), 'Trust and managerial problem solving', *Administrative Science Quarterly*, **17** (2), 229–39.

Zucker, L.G. (1986), 'Production of trust: institutional sources of economic structure', in B.M. Staw and L.L. Cummings (eds), *Research in Organizational Behaviour*, vol 8, Greenwich, CT: JAI Press, pp. 53–111.

3. Outsourcing for innovation: systems of innovation and the role of knowledge intermediaries

Jeremy Howells

INTRODUCTION

As the phenomenon of outsourcing has continued to expand, its impact is continuing to spread across a wide range of industries in terms of productivity growth, efficiency and innovation (Fixler and Siegel 1999; see also Heshmati 2003). This study focuses on the latter aspect, namely the role of outsourcing as it relates to innovative activity, and how this in turn relates to systems of innovation. Various terms have been applied to this activity, namely: 'research outsourcing', 'R&D outsourcing', 'research and technology outsourcing', 'scientific and technical knowledge outsourcing' and 'innovation outsourcing'. These terms may be divided into two basic forms: those focusing on inputs (knowledge, R&D and technology) that are sourced and those centred on outputs arising from such sourcing activities (innovation). In reality, most studies cover a range of issues associated with what might be termed 'sourcing for innovation'. There has also been a growing recognition that innovation is not just about specific R&D and technical requirements, but involves a much wider range of design, engineering and prototyping and testing activities and associated knowledge forms (from traditional forms of scientific knowledge through to market, financial and legal knowledge). This analysis uses the broad and more inclusive term 'outsourcing for innovation' to cover all these aspects of innovation outsourcing,[1] but also uses more narrow terms, such as research and technology outsourcing in the discussion.

This chapter is divided into three main sections. First, the chapter explores outsourcing processes in relation to innovation at a firm and organizational level (section 2). The chapter then moves on to a more general discussion of innovation outsourcing in terms of a systems of innovation approach (section 3). The last main section then focuses on one particular aspect of research and technology outsourcing, namely the growth of intermediation in this process by 'knowledge intermediaries' (section 4). The chapter ends with a short conclusion.

Before the chapter explores each of these issues in turn, and as a way of introduction, it is worth indicating the size of outsourcing activity, or at least parts of it, in relation to innovation. The expansion of extra-mural R&D within the UK economy can be seen as an exemplar of the growth of outsourcing activity associated with the innovation process more generally within advanced industrialized economies. The data reveal that by 2002, the amount spent by UK firms for R&D work in other UK firms was just under £1 billion (Table 3.1). Since 1995, the growth in this funding of research by other firms in UK firms has paralleled the growth in the funding of intra-mural research undertaken by companies in-house. A number of caveats should, however, be noted in relation to the data here. First, as noted above, R&D is only one input in the process of innovation; other significant elements in the innovation process include design, engineering, testing and trialling and other organizational innovative activities supporting them, such as market research. Second, the data shown below for the UK are not straightforward. The data indicate the value and proportion of funds that a firm receives from other firms to undertake R&D for them – so in effect the data reflect the size for research outsourcing, not by the firm outsourcing the work, but rather the firm undertaking the sourced R&D work. Last, the data are likely to underestimate R&D outsourcing in that they only record research funding by UK firms for work undertaken in other UK firms, not for R&D work spent overseas. They also do not record R&D funding by UK companies in non-firm organizations, such as universities, public research establishments and non-profit research organizations.

Table 3.1   Sources of funds for R&D undertaken in UK firms

	1995	1996	1997	1998	1999	2000	2001	2002
(a) Funding from other UK business	700	691	642	659	750	779	764	942
(b) Own funds	5 723	5 742	6 198	6 141	6 824	7 244	7 455	7 712
(a) as a % of (b)	12.23	12.03	10.36	10.73	10.99	10.75	10.25	12.21
(Total Business Enterprise R&D)	9 116	9 217	9 556	10 133	11 302	11 510	12 336	13 110

Source: compiled and adapted from Office for National Statistics (2003, 2004a, 2004b) data.

OUTSOURCING FOR INNOVATION AND THE FIRM

Introduction

Before examining some of the wider issues associated with outsourcing for innovation, it is important to analyse the sourcing process in relation to the firm itself. The remainder of this section will focus on four particular aspects. First, it will outline the distinctive features or peculiarities of outsourcing in relation to innovation as compared to other types of outsourcing activity. Second, it will look at the changing nature of outsourcing forms in relation to innovation, associated with a shift away from one-off contracts to ongoing multiple relationships. It is argued that the process has moved from uni-directional one-to-one relationships through to one-to-many relationships and on to interactive, multiple relationships. Linked with this trend, the analysis will then highlight the shift in innovation outsourcing strategy away from meeting more specific and shorter-term research and technology requirements for specific research or product programmes towards satisfying a firm's long-term requirements in terms of what technological capabilities and repertoires it needs to position itself for the future. Last, with the growth in multiple, interactive links, there is a discussion of how firms need to acknowledge and manage these complex, distributed network relationships.

The Distinctive Nature of Outsourcing for Innovation

There are important issues and strategies that firms need to be aware of in relation to outsourcing in general. However, outsourcing associated with innovative activity has a number of peculiarities which are important to note at the outset. Sourcing knowledge, research and technology inputs externally is different from other types of outsourcing activities in a number of ways:

1. The levels of risk and uncertainty in outcome involved are high (Doctor et al. 2001).
2. The issue of prior disclosure and information asymmetry about quality inherent in the market transfer of information are an essential feature of research and innovation outsourcing in that the client organization cannot determine the quality of the knowledge they receive unless the prior holder discloses it beforehand (Arrow 1962).
3. However, more especially the supplier of the knowledge often does not know the quality of the knowledge they sell in long-term research projects because they themselves do not know, a priori, the future

outcomes of their work. This characteristic, in turn, highlights the contractual incompleteness problem in relation to intellectual property exchange involved, where it is not possible to exhaustively stipulate the appropriate conditional responses in a contract.

4. A key feature of many knowledge intensive service activities, and particularly true in research and technology sourcing, is that both producer and consumer are involved to some extent in co-joint production of new knowledge (Alchian and Demsetz 1972), again leading to intellectual property rent-sharing issues (see below).

5. Related to this there are also a whole set of related moral hazard problems: (a) in that client organization can use the existing knowledge in a way which is not easily observable by the supplier (de Laat 1999, p. 209); (b) in that the supplier can use the existing knowledge provided to the recipient to other potential partners and which is not easily observable by the client organization; (c) in that the supplier can use co-produced knowledge, generated in conjunction with the recipient, and supply it (intentionally or unintentionally) to another client organization.

6. Research and technology are a central part of the future core competences and capabilities of the firm (Ford and Farmer 1986).

7. The often irreversible nature of the outsourcing decision in terms of its effect on R&D or technical capacity, at least over the short- and medium-term (see Welch and Nayak 1992).

8. The exchange of information associated with knowledge and research is a unique event, and is therefore not like other transactions which are repeatable in the nature of what is exchanged (Carter 1989, p. 157). Learning about the process is therefore more limited in nature.

9. Much of the know-how exchanged is often highly tacit in nature (Howells 1996; Cavusgil et al. 2003) but this is often difficult to control or formally administer within a contract.

Some of these features and issues can be seen in other outsourcing activities, particularly associated with other knowledge intensive collaborations and transactions related to IT projects (Miozzo and Grimshaw, Chapter 4 this volume). However, the inter-linked nature of the issues noted above, together with the dominant features of (a) high knowledge component combined with (b) the unknown outcomes of many R&D activities involved in the transaction, means that outsourcing for innovation has particularly distinctive features.

The issue of the 'unknown' is crucial here, because so much of the outcomes of research and other forms of innovation outsourcing is uncertain and this colours all aspects of the sourcing relationship. The issue of the prior disclosure problem (point 2 above) associated with knowledge exchange is well known and has long been reiterated in relation to R&D outsourcing, but this is about the exchange of existing knowledge. The exchange of existing technological knowledge in relation to technology licensing is extremely important (Lowe and Taylor 1998, p. 264), but the whole process of exchange or sourcing becomes even more complex when we discuss the issue in relation to new knowledge. Here both the receiver and the supplier do not know the final quality of the knowledge they are exchanging (point 3).

From Single, One-off Contracts to Multiple, Ongoing Relations

The distinctive nature of outsourcing for innovation makes the sourcing process a complex and uncertain process for the firm. However, there have been other trends which have further increased the complexity of outsourcing practices.

It should be noted that there are a variety of different forms of research and technology outsourcing. They range from short-term, one-off, contract-based links where the outcomes are already known or fairly certain and predictable to ongoing longer-term relations where outcomes are often highly uncertain and risky. The types of collaborations which are likely to be mainly contract-based are either in terms of licensing in existing technologies or through specific, short-term research or technical contracts. The most immediate and short-term source of external technology is via licensing in specific technologies (Atuahene-Gima 1992; Lowe and Taylor 1998). Companies often prefer research contracts to be short-term in nature, as long-term contracts are unsatisfactory inasmuch as most of the relevant contingences cannot be delineated (Teece 1976, p. 13). These types of links have been defined in more general form by Ring and van de Ven (1992, p. 487) as 'recurrent contracts' which involve repeated exchanges and their duration is relatively short-term. However, Ring and van de Ven (1992) have also identified what they define as 'relational contracts' which are associated with less formal types of relationship. These relational contracts are seen as involving longer-term investments where the partners actually do accept that outcomes cannot be fully specified in advance and frameworks are provided for future negotiations according to a set of rules. These types of longer-term relationships in innovation outsourcing appear to have become much more common (Howells et al. 2003) as firms seek deeper involvement and commitment from partners.

These changing relationships in outsourcing arrangements have, in turn, been part of a much wider distributed process of network relations as firms shift away from relatively rare one-off events on a one-to-one basis with an

R&D or technical supplier to ongoing multiple and parallel relationships associated with a range of different suppliers. This has had a number of other implications for the process and study of outsourcing for innovation:

1. First, it is now much harder to distinguish an outsourcing 'event' as outsourcing relationships move from a specific project or contract being outsourced to ongoing relational links. This has conceptual and methodological implications in that it is becoming ever harder to see (and measure) changes as a progression of snapshots of different time periods or 'rounds' (n, $n+1$, $n+2$, etc.) as opposed to an ongoing and never-ending change process, where one round ends and another starts is becoming impossible to distinguish.
2. Second, there is the issue of the changing nature of the 'source'. With the shift towards networks in innovation outsourcing, there is usually no longer a single source of knowledge or innovation input. Rather, multiple sources of knowledge are emerging, and are required, for any one solution in a research programme or new product development.
3. Third, and related to the above, with much closer interaction in innovation sourcing the distinction between supplier and recipient becomes less clear. In any one sourcing event (if this can be discerned, as noted above), a 'recipient' will also be providing information and knowledge to the 'supplier'. Bi-polar distinctions between the two types will give way to more relatively 'predominant' roles, but often as with knowledge intensive service exchanges and collaborations there is a high degree of knowledge co-production. Increasingly large manufacturing firms, such as Nortel, which source-in R&D and technical inputs also operate units, which provide R&D outsourcing services to other firms.

From Sourcing Inputs to Creating Capabilities

Accompanying these changes, there has also been a shift in what firms are seeking from research and technology sourcing. Firms have become more strategic about their sourcing requirements for innovation in terms of the timescale they view their requirements and what the nature of their requirements that need satisfying are.[2] Thus firms have moved from simply being concerned about meeting their immediate innovation requirements (spanning perhaps two to three years) towards setting out what they might need in the longer term (spanning from five to ten years). This relative shift has been associated with a move from focusing on direct inputs into a problem or outcome to seeking external capabilities (to complement internal capabilities) which are needed to create future inputs for projects and programmes.

These former more immediate requirements associated with short-run knowledge and technology relations would be centred on 'problem-oriented' innovation and the objectives for such links would be on specific 'sub-unit' (not organization-wide) goals (see Cyert and March 1963, p. 279). By contrast, the latter focus would be centred on gaining particular knowledge and technologies that would be used by the firm to augment its current technology base and providing additional 'external capabilities' (Langlois 1997) which can be deployed by the firm. These external sources in combination with the in-house generation of new knowledge are then deployed to produce new and improved products and processes, which are hopefully aligned with future market requirements.

This longer strategic capabilities perspective of the firm associated with innovation outsourcing is having a profound influence on how the firm uses long-term knowledge and technology relations with other firms and organizations. The main objective of these type of arrangements is to help align the future technological competence of the firm with future market conditions, so that the future dynamic capabilities of the firm (Elfring and Baven 1994) are in a good position to deliver future new products and services to satisfy the market and to be competitive in this respect against other existing and potential firms. It should be noted that it is just as important for the firm to identify technologies and markets not to move into and satisfy, as much as selecting those to enter and develop. If the 'knowledge base' of the firm can be narrowly defined as the specific technologies and markets of which it has experience (Metcalfe and de Liso 2000), then the focus here is on the firm positioning itself in relation to what it wants as its future 'knowledge' base.

These long-term objectives may seem relatively straightforward, but they are highly complex and difficult to ascertain and achieve. To be successful, not only does the firm have to predict future technological and market conditions fairly accurately, but with regard to its technological base, the firm has to decide what technologies it should provide internally and what it requires from external sources. Underlying this technology framework, the company needs to position its knowledge requirements, both in internal knowledge generation and external knowledge acquisition, to meet these technological needs. This is centred on the interplay between the nexus of the capability of the firm to exploit its knowledge and the unexplored potential of the technology (Kogut and Zander 1992, pp. 391–2).

Managing Innovation Outsourcing Within Distributed Networks

If we acknowledge that sourcing for innovation has become more centrally involved with developing strategic capabilities, it has also become involved with managing multiple technology relations (see above). Increasingly, the innovation process can be characterized as 'distributed'. Innovation requires the combination of capabilities distributed amongst business units within and between firms (Coombs and Metcalfe 2000; Malik 2003). As noted above, the selection and use of external knowledge and technology sources can no longer be viewed on a one-to-one basis, but has to be considered and managed on a one-to-many basis in acknowledgement of the fact that a firm has to collaborate with many different partners within a network or more informal cluster of firms and organizations, i.e. within its distributed innovation community. This recognition within the resource-based literature that firm-addressable knowledge assets can be selectively traded between firms and that they can play an important role in the innovation process has led to a growing interest in the way that networks of organizations are created and evolve over time (Madhavan and Grover 1998). Increasingly, it is argued that a firm's performance is directly linked to its effort at such competence-building and renewal and the creation and development of networks linked to the innovation process are regarded as a potential source of competitive advantage. The management of innovation sourcing is therefore not just about managing single sourcing but also about managing multiple network relations and how a firm positions itself within these networks. Firms are having to increasingly accept that rather than narrowly considering the relative merits of a (bilateral) technology source in relation to its short-term requirements, it has to take a long-term view of its technological requirements within an (multilateral) innovation community setting, or indeed with different sets of innovation communities.

The increasingly distributed nature of innovation therefore places a much greater 'architectural' (configurational) burden on firms when considering future technological commitments. The potential source or coupling point of the newly emerging technology and knowledge base has to be considered not only in terms of its direct potential for achieving technological objectives, but also the likely commercial and market contexts for that technology and the network relationships of the technology supplier. It has already been acknowledged that future technologies may be successful in meeting their technological requirements but for various reasons, may not then be commercially successful. An increasingly important factor in commercial success is now network-dependent, either in terms of being part of the actual or de facto industry standard or in an innovation community that becomes the dominant design. Selecting a potential technology or knowledge route can have major strategic implications if it places a firm in a 'non-winning' innovation community or network. In mobile telecommunications, certain supplier companies in relation to technologies or components, have arguably faired much better with being in the Nokia innovation community than those of Ericsson, Motorola or Philips, although the technologies they have delivered or acquired have been equal or superior in technology and design terms.

The distributed nature of innovation means that firms are therefore becoming more inter-dependent for successful outcomes in their techno-logical routing. By being a member of an innovation network, in one sense it can be said to lower the risks of technological failure, as the burden for exploiting the new technology is no longer borne by one firm. However, in another sense, risks are increased for firms as they not only have to evaluate their own capability of being successful, but also that of their partners. In addition, except for a few major, lead companies, most firms now can there-fore be said to be in less control of their technological destinies than they were. Risks of failing in the technology may now be subsumed under the wider risk of not being in the right network or community. Staying in the technological race may now be less about satisfying the specific technologi-cal requirements of the project, and more about successfully managing the requirements of the wider network (Bailetti et al. 1998) and these require-ments may be much more burdensome. Being able to manage sustained cooperation with others, especially in difficult or novel ventures (Ring and van de Ven 1992, p. 487) has therefore become an increasingly important managerial asset for firms to develop but one that is rarely considered.

By seeking to deliver a certain new technology (which in turn then requires external knowledge and technology inputs from other partners) within a newly emerging innovation community can therefore have a long-term impact in relation to the future shaping of a firm's technological profile. The partners and the network may be just as important in deter-mining the success of the technology as the technology per se. The position of the firm within the network or community, and what network is selected are all critical decisions here. Selecting or helping to create innovation com-munities, sustaining them and refreshing them (as old members may be cast out and new ones introduced; Lynn et al. 1996, p. 100) and managing the relations between the partners and the subsequent outputs in relation to intellectual property all become important here for managers to learn and develop.

OUTSOURCING FOR INNOVATION AND INNOVATION SYSTEMS

The growth of research and technology outsourcing has a number of important conceptual and methodological issues for the systems of innov-ation approach. Can we, for example, conceive innovation outsourcing networks and communities within a system of innovation framework? If so, how do innovation systems fit within larger (national or sectoral) systems of innovation frameworks? There are a number of dimensions to be considered here. These centre on: definitional, boundary and nesting issues; connectivity and openness within and between systems; temporality; evolutionary and dynamic aspects of the system; and, the basic elements of the system (the changing role of firms and agents (nodes) and their changing relationships (links)). Each of these will be briefly explored in turn as they relate to a system of innovation approaches.

1) A key element in defining any system is being able to draw a boundary around the set of elements (firms and organizations) and their chosen links within it. Delimiting an innovation outsourcing system obviously cuts across many technological systems and national systems of innovation which have already been highlighted in earlier studies. The key elements in such a system would be both research and technical service companies and their 'manufacturing' (but also service) recipients (forming a wider 'organizational population' of different actors and agencies; Reddy and Roa 1990; see also Phillips 1960). The boundary of the system could be defined in straightforward 'market' terms, but as has been indicated above, there are strong informal and non-traded elements in the innovation outsourcing 'market' which involve high levels of reciprocity and trust and 'relational' networks (Dore 1986). In terms of this wider system, many of the links described would not be based on purely monetary transactions and formal contracts, but involve informal know-how trading and patterns of exchange (see, for example, Von Hippel 1987; Kreiner and Schultz 1993). Few systems have clear and mutually exclusive elements, links and boundaries, but in particular with innovation outsourcing systems there would be a considerable overlap and nesting of different systems. There is little acknowledgement in systems of innovation literature that firms or organizations can be members of multiple innovation systems. This represents a key danger of a systems approach in that it encourages the conceptual mindset of having separate and 'closed' (in terms of not having, or rather, not considering links and interrelationships external to the system) innovation systems.

2) The growth of innovation outsourcing and externalization will also more generally lead to an increase in the overall connectivity within and openness between innovation systems. On a more abstract level, and in terms of systems theory, this may suggest innovation systems have increased susceptibility to volatility, dislocation and radical change, but equally it might suggest that by allowing greater diversity of connections it promotes greater stability and choice (Borg 2001, p. 519). The evolution of the system will also have an impact on the spatial and temporal diffusion

processes in relation to innovation. With increased connectivity and inter-linkage, innovation waves will diffuse in significantly different ways. Above all, innovations will diffuse more rapidly throughout the system associated with increased and more pervasive information flows between diffusion actors (although major barriers will remain; see, for example, Turpin et al. 1996, p. 271).

3) Leading on from this is that although innovation systems literature has so frequently stressed the temporal dynamic qualities of such systems and the approach, it still goes on to treat such systems in a static fashion. The wider innovation outsourcing system, or one of its sub-systems, in period $n1$ is going to be different in period $n2$. Key elements or actors will have disappeared between the two periods and have been replaced, leading to new and different linkage patterns and changes in the overall shape and configuration of the system (with some of these changes being measurable and therefore allowing systems to be compared). Much of what a system is, depends on what phenomenon (or what attribute of a phenomenon) we wish to study. This determines what the 'element' (or basic unit) within the system will be. It may be best to treat the sourcing of innovation as a 'phenomenon-based' system of innovation which can be seen as nesting and overlapping with existing geographical and sectoral systems of innovation, but in a more time-specific and dynamic set-up. The transitory and evolutionary nature of such systems in this way is stressed but it accepts that other, more macro national or sectoral systems are operating at the same time and in part provide part of the institutional, economic and social context for these meta systems. Indeed, this perspective parallels Lynn's et al. (1996, pp. 99–101) notion of an innovation community which stresses the dynamic elements of change within such communities, such as entry and exits, as well as methodological issues to do with boundary permeability and connectedness of organizations within a group.

4) Over the longer term, however, there is the issue of how the system evolves. This evolution is strongly influenced by institutional, as well as more general cultural, social and economic, elements. Institutions are central to the notion of any kind of system of innovation (Edquist 1997, p. 24). They provide the dynamic context (the routines and 'guide posts'; Andersen and Lundvall 1997, pp. 248–51; Lundvall 1992, p. 10) through which systems evolve and change, linked with the notion of technological paradigms and trajectories. However, whilst the creation and development of external networks may well be a potential source of competitive advantage, past work has tended to treat such networks as givens and there has been little attention paid to how networks are created or how they evolve over time. Thus, there has been little discussion about how systems emerge. In relation to outsourcing for innovation on the contracting side, markets

for research and technology services are emerging which are still very fragile and nascent in their formation (Howells 1999a), but there is only fragmentary understanding about how these embryonic markets are formed or instituted during these early phases. At the same time, there has been relatively little attention paid to the reasons why networks change over time, the factors that drive their evolution or the impact of specific industry or firm-level events (Madhavan and Grover 1998). Nevertheless, there is some evidence to suggest that emerging firms in new sectors purposefully create firm-specific innovation systems (i.e., networks of external relationships that support the innovative activities of the firm) and that those firm-specific networks evolve over time in response to the changing requirements of the firm, the maturing of markets and the changing demands of the innovation process (Metcalfe and James 2000).

5) As noted earlier, the nature of the nodes and links within the innovation outsourcing system are changing and becoming blurred. Firms are no longer just suppliers or recipients of knowledge and technology, they are becoming both. Equally, dyadic inter-organizational relationships are giving way to more simultaneous and parallel relationships. The shift from serial to simultaneous and parallel working has therefore become more commonplace. Such multi-collaborations are centred around a particular technology or product, but in large, complex corporations dealing with a whole series of technologies and products such networks become increasingly complex. However, existing studies still tend to focus on single types of external collaboration: contract research, technology licensing, alliances and joint ventures or more general cross-institutional associations: industry–academic collaboration; industry–public sector research links; and industry–industry technology contact. As such these studies have concentrated mainly on *object-based* analysis of innovation (Archibugi and Michie 1995), rather than on a more specific *subject-based* approach of companies and their complex web of links; especially absent is a more holistic view of technological links by a specific firm.

THE ROLE OF KNOWLEDGE INTERMEDIARIES IN INNOVATION

The role of knowledge intermediaries within the innovation process can be linked to the growth of knowledge intensive business services (KIBS; Bettencourt et al. 2002; Miles et al. 1995; Miles 2000; O'Farrell and Moffat 1991; O'Farrell and Wood 1999). Many KIBS firms have close and continuous interactions with their clients which can involve crucial, but largely hidden, functions in supporting innovative change within their client companies (Wood 2002, p. 997; see also Bessant and Rush 1995). In turn, this has also been recognized in the increasing role of KIBS organizations in the wider innovation system (Muller and Zenker 2001).

The analysis of research and technology outsourcing in relation to innovation systems also has important implications for how service firms, activities and indeed innovations are viewed in existing literature. As noted elsewhere (Howells 1999a, 1999b; Howells and Roberts 2000), manufacturing firms are still seen as taking the 'centre stage' with tight systems defined around them. By contrast, service firms have been viewed as being at the edges of the system and/or operating in a supporting or indirect role (as subsidiary or second-tier 'elements') within the innovation system.

Service firms and organizations (including contract research organizations, design and engineering consultancies, public research establishments and so on) which provide many of the research and technology services for these manufacturing firms are therefore seen as representing the passive 'supporting cast'. They at best represent what are acknowledged as key elements in the innovation infrastructure (see, for example, Smith 1997), but their role is still firmly placed as 'supporters' rather than those who lead the innovation process. Admittedly, this is in part because of a slow institutionalization process for these contract, engineering, technical and design firms to emerge and form clearly identifiable and separate service sectors (see Walsh 1996, pp. 523–4, for example, in relation to design services). However, this 'received' view is increasingly being challenged as contract research organizations and engineering and design consultancies take a more proactive role, often initiating contact with the manufacturing firm (Howells 2000; von Emloh et al. 1994). Such companies not infrequently take the 'lead' in the innovation process and undertake their own R&D work. They commission the research and undertake much of the basic groundwork R&D before then approaching likely manufacturing firms who may participate in delivering support manufacturing and technical inputs to produce an innovation. Manufacturing firms represent the passive elements here, whilst the service firms are the *proactive* elements and represent the 'innovation integrators' and leaders. In the Cambridge area, for example, companies, such as ARM, Generics, PA Consulting and TTP Communications, undertake no manufacturing, but play a pivotal lead role in the design, innovation and new product development process. Such companies initiate developments and then license out their technologies to, for example, chip manufacturers and mobile phone producers or enter into long-term research, design and development contracts with them.

Many of these more proactive strategies have indeed grown up around one-to-one relationships with key manufacturing (and service) companies. However, as the relationships have become more complex within wider 'loosely coupled' systems (Sundbo and Gallouj 2000) the simple dichotomous 'supplier' and 'recipient' role classification in these networks has given way to more complex and varied roles. What position or role a firm may take within a network may vary not only between projects and over time but even for the same potential project (Howells 1999b). A firm A bidding for a project, may in one tender proposal be a collaborator with another firm B in one network coalition of partners, but at the same time be a competitor with firm B in a different coalition of partners in another tender proposal for the same project.

Within this more complex realm has emerged a set of actors who might be termed 'knowledge intermediaries' who perform the role of 'third parties' (Mantel and Rosegger 1987), 'bridging organizations' (Carlsson 1995, p. 444; see also Carlsson and Stankiewicz 1991), 'bricoleurs' (Turpin et al. 1996, p. 271; see Levi-Strauss 1966) 'brokers' (Provan and Human 1999) or 'intermediaries' (specifically associated with information exchange; Popp 2000) between single supplier and recipient relationships, but also between multiple supplier and recipient networks and *within* such networks. The role of such intermediaries is not just confined to inter-firm relations, but also between firms and universities and other public and quasi-public research organizations (Shohert and Prevezer 1996, p. 291). The role of intermediary can also be traced back to much earlier forms in the shape of, for example, middlemen in the wool and textile industries of sixteenth, seventeenth and eighteenth century Britain (Farnie 1979; Hill 1967).

One such example is that of CERAM, an independent research and technology organization involved in ceramics technology and based in Stoke-on-Trent in the UK. CERAM grew out of being the industrial research association for the refactories and pottery industries which had separate research associations and merged in 1946 to form the British Ceramic Research Association (Johnson 1973). CERAM has moved from being a relatively *reactive* contract research, testing, analysis and standards provider for its members in the UK to a much more *proactive*, global research and technology institute. CERAM, for example, became aware that a large multinational retail company with extensive global operations was being supplied by low-cost ceramics manufacturers in central and eastern Europe had a very high fracture and breakage rate. The company had considerable and long-term experience with the furniture and wood industry, but very little experience of the pottery and ceramics industries. CERAM therefore put forward a proposal to the company to improve their suppliers' performance in terms of reducing breakage and wastage rates through the use of computational modelling, process and tunnel kiln simulation.[3] Working with the ceramic factories across central and eastern Europe, CERAM is managing to accurately predict the thermal and mechanical performance of the pottery being produced and also reducing the quantity of experimental trials required before full-scale production. Through this work, CERAM is managing to reduce very high wastage rates in the suppliers' factories, and is also providing an 'intelligent customer' role for the company in the future commissioning and purchasing of ceramics for its stores. It is, therefore, not only working with both suppliers and recipients, it is also providing immediate 'recurrent' research services to the suppliers in partnership with the company, but also providing longer-term 'relational' research capabilities for it.

Another example is that of LGC, based in London and Runcorn, which until 1996 was formerly a unit of the UK government department, the Department of Trade and Industry. LGC still has an official role of 'Government Chemist' providing independent advice on reference, statutory and regulatory issues. It has, however, emerged as a major analytical service provider to companies and organizations, in a wide range of industries (outside its home base in the chemicals industry). In addition, it is seeking to move away from being solely a reactive high-volume, low-margin supplier of analytical services to a large number of customers to becoming a more proactive supplier providing complete solutions to key client companies. Thus, LGC has taken over the responsibility of all the analytical activities for a major UK chemical company, with LGC staff working on-site within the client firm. This role covers not only day-to-day, problem-solving activities, but also a more strategic intermediary role by becoming an 'intelligent interface' between its client and its 'task environment' in relation to analytical, environmental and testing matters. This includes: providing advice on what the client company should be doing in the future with regard to analytical activities; how it should react to the changing regulatory environment; provide hazard assessments; outline what improvements can be made in relation to measurement and testing techniques and so on.

A third and final example is that of a Cambridge-based consultancy company, Oaklands. One of its clients was a major, process-based manufacturing company and it provided advice on what external research links (especially in relation to university links) the company would find valuable for its long-term development. Again, this was not just a simple scoping, inventory type exercise of finding potentially useful research expertise in universities and other research centres. It also involved undertaking a deeper understanding of what the client company actually needed, identifying what the client companies' core competences were that were important for its long-term success and then mapping potentially useful research links with this profile. This centred on identifying where external researchers could fill the current and future research and technical gaps (or 'weak areas') that the client company could not provide or would do better not providing itself.

Oaklands has now moved further by not simply identifying research gaps and finding suitable partners to fill these gaps, but by moving more directly towards a more high-level operation of finding research and technical solutions to the client company's perceived short- and long-term problems.

CONCLUSIONS

This review and analysis has tried to show that outsourcing for innovation is more than just a 'research or buy' decision. It involves a much wider range of knowledge intensive services than just R&D and these are often consumed together. It also encompasses much longer-term relationships centred around building research and knowledge capabilities which the firm needs in the future, as well as more immediate R&D and technical requirements. Last, outsourcing for innovation has moved from dyadic one-to-one linear relationships to much more interactive multiple links within distributed networks. All these trends have made sourcing for innovation a much more complex and blurred process to study. Contract research and technology organizations, design and engineering consultancies and other knowledge intermediaries, should no longer be seen as passive agents or 'supporting cast' in relation to innovative activity. Increasingly they are taking on more proactive, central and pivotal roles in the innovation process.

This chapter has also illustrated the changing roles and emergence of different actors within the innovation outsourcing arena. The roles of knowledge and research recipients and suppliers have become less distinct as firms and organizations become both suppliers and consumers of research and technical services. Equally, there has been the emergence of new forms of actors and agents such as knowledge intermediaries who appear to becoming more important within the overall innovation system and becoming pivotal facilitators in networks of research and technology co-production.

ACKNOWLEDGEMENT

Thanks go to all the participant case study interviews coming from ARM, CERAM, Generics, LGC Holdings, Oaklands, PA Consulting, TTP Communications and other AIRTO members. This chapter arises out of research funded by the ESRC (Grant Number ESRC L700377003) and ESRC, EPSRC and Link Programme (Grant Number L700257003) research projects. The views expressed are the author's alone.

NOTES

1. The term 'innovation outsourcing' is somewhat misleading, since it infers the whole process of innovation, including commissioning and project management, to be outsourced to another organization, which is very unlikely.
2. However, operational issues may still remain important in some forms of consumption and outsourcing (see Chen 1997, p. 134).
3. In turn involving Finite Element Analysis (FEA), a form of computational modelling, Factorial Experimental Design (FED), and Computer Aided Design (CAD) and Computer Aided Manufacturing (CAM).

REFERENCES

Alchian, A.A. and H. Demsetz (1972), 'Production, information costs and economic organization', *American Economic Review*, **62** (5), 777–95.

Andersen, E.S. and B-Å. Lundvall (1997), 'National innovation systems and the dynamics of the division of labor', in C. Edquist (ed.), *Systems of Innovation: Technologies, Institutions and Organizations*, London: Pinter, pp. 242–65.

Archibugi, D. and J. Michie (1995), 'The globalisation of technology: a new taxonomy', *Cambridge Journal of Economics*, **19** (1), 155–74.

Arora, A. and A. Gambardella (1990), 'Complimentarity and external linkages: the strategies of the large firms in biotechnology', *Journal of Industrial Economics*, **38** (4), 361–79.

Arrow, K.J. (1962), 'Economic welfare and the allocation of resources for inventions', in R. Nelson (ed.), *The Rate and Direction of Inventive Activity*, Princeton: Princeton: NJ University Press, pp. 609–25.

Atuahene-Gima, K. (1992), ' Inward technology licensing as an alternative to internal R&D in new product development: a conceptual framework', *Journal of Product Innovation Management*, **9** (2), 156–67.

Atuahene-Gima, K. and P. Patterson (1993), 'Managerial perceptions of technology licensing as an alternative to internal R&D in new product development: an empirical investigation', *R&D Management*, **23** (4), 327–36.

Bailetti, A.J., J.R. Callahan and S. McCluskey (1998), 'Coordination at different stages of the product design process', *R&D Management*, **28** (4), 237–47.

Bessant, J. and H. Rush (1995), 'Building bridges for innovation: the role of consultants in technology transfer', *Research Policy*, **24** (1), 97–114.

Bettencourt, L.A., A.L. Ostrom, S.W. Brown and R.J. Roundtree (2002), 'Client co-production in knowledge-intensive business services', *California Management Review*, **44** (4), 100–28.

Borg, E.A. (2001), 'Knowledge, information and intellectual property: implications for marketing relationships', *Technovation*, **21** (8), 515–24.

Carlsson, B. (1995), 'The technological system for factory automation: an international comparison', in B. Carlsson (ed.), *Technological Systems and Economic Performance: The Case of Factory Automation*, Dordrecht: Kluwer, pp. 441–75.

Carlsson, B. and R. Stankiewicz (1991), 'On the nature, function and composition of technological systems', *Journal of Evolutionary Economics*, **1** (2), 93–118.

Carter, A.P. (1989), 'Knowhow trading as economic exchange', *Research Policy*, **18** (3), 155–63.

Cavusgil, S.T., R.J. Calantone and Y. Zhao (2003), 'Tacit knowledge transfer and firm innovation capability', *Journal of Business and Industrial Marketing*, **18** (1), 6–21.

Chatterji, D. (1996), 'Accessing external sources of technology', *Research Technology Management*, **39** (2), 48–56.

Chen, S-H. (1997), 'Decision-making in research and development collaboration', *Research Policy*, **26** (1), 121–35.

Chesbrough, H. and D.J. Teece (1996), 'When is virtual virtuous? Organizing for innovation', *Harvard Business Review*, **74** (1), 65–73.

Chiesa, V. and R. Manzini (1997), 'Managing virtual R&D organizations: lessons from the pharmaceutical industry', *International Journal of Technology Management* **13** (5–6), 471–85.

Cook, P.L. (1987), 'Research and development networks and markets in a complex industry: the example of offshore oil equipment', in C.T. Saunders (ed.), *Industrial Policies and Structural Change*, London: Macmillan, pp. 105–17.

Coombs, R. and S. Metcalfe (2000), 'Organizing for innovation: co-ordinating distributed innovation capabilities', in N. Foss and V. Mahnke (eds), *Competence, Governance, and Entrepreneurship: Advances in Economic Strategy Research*, Oxford: Oxford University Press, pp. 209–31.

Cyert, R.M. and J.G. March (1963), *A Behavioural Theory of the Firm*, Englewood Cliffs, NJ: Prentice Hall.

De Laat, P.B. (1999), 'Dangerous liaisons: sharing knowledge within research and development alliances', in A. Grandori (ed.), *Interfirm Networks: Organization and Industrial Competitiveness*, Routledge, London, pp. 208–33.

Doctor, R.N., D.P. Newton and A. Pearson (2001), 'Managing uncertainty in research and development', *Technovation*, **21** (2), 79–90.

Dore, R. (1986), *Flexible Rigidities: Industrial Policy and Structural Adjustment in the Japanese Economy: 1970–1980*, London: Athlone Press.

Edquist, C. (1997), 'Systems of innovation approaches – their emergence and characteristics', in C. Edquist (ed.), *Systems of Innovation: Technologies, Institutions and Organizations*, London: Pinter, pp. 1–35.

Elfring, T. and G. Baven (1994), 'Outsourcing technical services: stages of development', *Long Range Planning*, **27** (5), 42–51.

Farnie, D.A. (1979), *The English Cotton Industry and the World Market, 1815–1896*, Oxford: Oxford University Press.

Fixler, D. and D. Siegel (1999), 'Outsourcing and productivity growth in services', *Structural Change and Economic Dynamics*, **10** (2), 177–94.

Ford, D. and D. Farmer (1986), 'Make or buy – a key strategic issue', *Long Range Planning*, **19** (5), 54–62.

Håkansson, H. (ed.) (1987), *Industrial Technological Development – A Network Approach*, London: Croom Helm.

Haour, G. (1992) 'Stretching the knowledge-base of the enterprise through contract research', *R&D Management*, **22** (2), 177–82.

Häusler, J., H-W. Hohn and S. Lütz (1994), 'Contingencies of innovative networks: a case study of successful interfirm R&D collaboration', *Research Policy*, **23** (1), 47–66.

Heshmati, A. (2003), 'Productivity growth, efficiency and outsourcing in manufacturing and service industries', *Journal of Economic Surveys*, **17** (1), 79–112.

Hill, C. (1967), *Reformation to Industrial Revolution*, London: Weidenfeld & Nicolson.

Howells, J. (1996), 'Tacit knowledge, innovation and technology transfer', *Technology Analysis and Strategic Management*, **8** (2), 91–106.

Howells, J. (1999a), 'Research and technology outsourcing', *Technology Analysis and Strategic Management*, **11** (1), 17–29.

Howells, J. (1999b), 'Research and technology outsourcing and innovation systems: an exploratory analysis', *Industry and Innovation*, **6** (1), 111–29.

Howells, J. (2000), 'Outsourcing novelty: the externalisation of innovative activity', in B. Andersen, J. Howells, R. Hull, I. Miles and J. Roberts (eds), *Knowledge and Innovation in the New Service Economy*, Cheltenham, UK and Northampton, MA, USA: Edward Elgar, pp. 196–214.

Howells, J. and J. Roberts (2000), 'Global knowledge systems in a service economy', in B. Andersen, J. Howells, R. Hull, I. Miles and J. Roberts (eds), *Knowledge and Innovation in the New Service Economy*, Cheltenham, UK and Northampton, MA, USA: Edward Elgar, pp. 255–74.

Howells, J., A. James and M. Khaleel (2003), 'The sourcing of technological knowledge: distributed innovation processes and dynamic change', *R&D Management*, **33** (4), 395–409.

Johnson, P.S. (1973), *Co-operative Research in Industry: An Economic Study*, London: Martin Robertson.

Kogut, B. and U. Zander (1992), 'Knowledge of the firm: combinative capabilities and the replication of technology', *Organization Science*, **3** (3), 383–97.

Kreiner, K. and M. Schultz (1993), 'Informal collaboration in R&D: the formation of networks across organizations', *Organization Studies*, **14** (2), 189–209.

Langlois, R.N. (1997), 'Transaction cost economics in real time', in N.J. Foss (ed.), *Resources, Firms and Strategies*, Oxford: Oxford University Press, pp. 286–305.

Levi-Strauss, C. (1966), *The Savage Mind*, Chicago: University of Chicago Press.

Lowe, J. and P. Taylor (1998), 'R&D and technology purchase through licence agreements: complementary strategies and complementary assets', *R&D Management*, **28** (4), 263–78.

Lundvall, B.-Å. (1992), 'Introduction', in B.-Å. Lundvall (ed.), *National Systems of Innovation: Towards a Theory of Innovation and Interactive Learning*, London: Pinter, pp. 1–19.

Lynn, L.H., N.M. Reddy and J.D. Aram (1996), 'Linking technology and institutions: the innovation community framework', *Research Policy*, **25** (1), 91–106.

Macneil, I. (1978), 'Contracts: adjustment of long-term economic relations under classical, neoclassical, and relational contract law', *Northwestern University Law Review*, **72** (6), 854–906.

Madhavan, R. and R. Grover (1998), 'From embedded knowledge to embodied knowledge: new product development as knowledge management', *Journal of Marketing*, **62** (4), 1–12.

Malik, K. (2003), 'Distributed capabilities: intra-firm technology transfer inside BICC Cables', *International Journal of Manufacturing Technology and Management*, **5** (4), 311–24.

Mantel, S.J. and G. Rosegger (1987), 'The role of third-parties in the diffusion of innovations: a survey', in R. Rothwell and J. Bessant (eds), *Innovation: Adaptation and Growth*, Amsterdam: Elsevier, pp. 123–34.

Metcalfe, J.S. and N. de Liso (2000), 'Innovation, capabilities and knowledge: the epistemic connection', in R. Coombs, K. Green, A. Richards and V. Walsh (eds), *Technological Change and Organization*, Cheltenham, UK and Northampton, MA, USA: Edward Elgar, pp. 8–27.

Metcalfe, J.S. and A. James (2000), 'Knowledge and capabilities – a new view of the firm', in N.J. Foss and P.L. Robertson (eds), *Resources, Technology and Strategy – Explorations in the Resource-Based Perspective*, London: Routledge, pp. 31–52.

Miles, I. (2000), 'Services innovation: coming of age in the knowledge-based economy', *International Journal of Innovation Management*, **4** (4), 371–89.

Miles, I., N. Kastrinos, K. Flanagan, R. Bilderbeek, P. den Hertog, W. Huntink and M. Bouman (1995), 'Knowledge-intensive business services: users, carriers and sources of innovation', *EIMS Publication*, **15**, published by the Innovation Programme, Directorate General for Telecommunications, Information Market and Exploitation of Research, Commission of the European Communities, Luxembourg.

Muller, E. and A. Zenker (2001), 'Business services as actors of knowledge transformation: the role of KIBS in regional and national innovation systems', *Research Policy*, **30** (9), 1501–16.

O'Farrell, P.N. and L.A.R. Moffat (1991), 'An interaction model of business service production and consumption', *British Journal of Management*, **2** (4), 205–21.

O'Farrell, P.N. and P.A. Wood (1999), 'Formation of strategic alliances in business services: towards a new client-oriented conceptual framework', *Service Industries Journal*, **19** (1), 133–51.

Office for National Statistics (2003), *Business Enterprise Research and Development, 2002*, London: National Statistics.

Office for National Statistics (2004a), *Gross Domestic Expenditure on Research and Development, 2002*, London: National Statistics.

Office for National Statistics (2004b), *Research and Experimental Development, (R&D) Statistics, 2002*, Newport: Office for National Statistics.

Phillips, A. (1960), 'A theory of interfirm organization', *Quarterly Journal of Economics*, **74** (4), 602–13.

Pisano, G.P. (1990), 'The R&D boundaries of the firm: an empirical analysis', *Administrative Science Quarterly*, **35** (1), 153–76.

Pisano, G.P. (1994), 'Knowledge, integration and the locus of learning: an empirical analysis of process development', *Strategic Management Journal*, **15** (special issue), 85–100.

Popp, A. (2000), 'Swamped in information but starved of data: information and intermediaries in clothing supply chains', *Supply Chain Management*, **5** (3), 151–61.

Powell, W.W. (1990), 'Neither market nor hierarchy: network forms of organization', in B.M. Staw and L.L. Cummings (eds), *Research in Organizational Behavior*, vol 12, Greenwich, CT: JAI Press, pp. 295–336.

Provan, K.G. and S.E. Human (1999), 'Organizational learning and the role of the network broker in small-firm manufacturing networks', in A. Grandori (ed.), *Interfirm Networks: Organization and Industrial Competitiveness*, London: Routledge, pp. 185–207.

Reddy, N.M. and M.V.H. Rao (1990), 'The industrial market as an interfirm organization', *Journal of Management Studies*, **27** (1), 43–59.

Ring, P.S. and A.H. van de Ven (1992), 'Structuring cooperative relationships between organizations', *Strategic Management Journal*, **13** (7), 483–98.

Robins, J.A. (1987), 'Organizational economics: notes on the use of transaction-cost theory in the study of organizations', *Administrative Science Quarterly*, **32** (1), 68–86.

Sen, F. and A.H. Rubenstein (1990), 'An exploration of factors affecting the integration of in-house R&D with external technology acquisition strategies of a firm', *IEEE Transactions on Engineering Management*, **37** (4), 246–58.

Shohert, S. and M. Prevezer (1996), 'UK biotechnology: institutional linkages, technology transfer and the role of intermediaries', *R&D Management*, **26** (3), 283–98.

Smith, K. (1997), 'Economic infrastructures and innovation systems', in C. Edquist (ed.), *Systems of Innovation: Technologies, Institutions and Organizations*, London: Pinter, pp. 86–106.

Steensma, H.K. and K.G. Corley (2001), 'Organizational context as a moderator of theories on firm boundaries for technology sourcing', *Academy of Management Journal*, **44** (2), 271–91.

Sundbo, J. and F. Gallouj (2000), 'Innovation as a loosely coupled system in services', *International Journal of Services Technology and Management*, **1** (1), 15–36.

Teece, D.J. (1976), *Vertical Integration and Vertical Divestiture in the U.S. Oil Industry: Analysis and Policy Implications*, Stanford, CA: Stanford University Institute for Energy Studies.

Turpin, T., S. Garrett-Jones and N. Rankin (1996), 'Bricoleurs and boundary riders: managing basic research and innovation knowledge networks', *R&D Management*, **26** (3), 267–82.

Tyler, B.B. and H.K. Steensma (1995), 'Evaluating technological collaborative opportunities: a cognitive modelling perspective', *Strategic Management Journal*, **16** (special issue), 43–70.

Varma, R. (1993), 'Restructuring corporate R&D: from an autonomous to a linkage model', *Technology Analysis and Strategic Management*, **7** (2), 231–45.

Veugelers, R. (1997), 'Internal R&D expenditures and external technology sourcing', *Research Policy*, **26** (3), 303–15.

von Emloh, D.A., A.W. Pearson and D.F. Ball (1994), 'The role of R&D in international process plant contracting', *International Journal of Technology Management*, **9** (1), 61–76.

Von Hippel, E. (1987), 'Cooperation between rivals: informal know-how trading', *Research Policy*, **16** (6), 291–302.

Walsh, V. (1996), 'Design, innovation and boundaries of the firm', *Research Policy*, **25** (4), 509–29.

Welch, J.A. and P.R. Nayak (1992), 'Strategic sourcing: a progressive approach to the make-or-buy decision', *Academy of Management Executive*, **6** (1), 23–31.

Wood, P.A. (2002), *Consultancy and Innovation: The Business Service Revolution in Europe*, London: Routledge.

4. Modularity and innovation in knowledge intensive business services: IT outsourcing in Germany and the UK

Marcela Miozzo and Damian Grimshaw

INTRODUCTION

Although IT outsourcing has been an accepted business practice since the 1980s, it has shown remarkable growth in recent years and has been the engine of growth for the software and computer services sector. Its nature, however, has changed dramatically. IT outsourcing has matured from a commodity service to 'risk/rewards' partnerships. In several countries (such as the UK, the USA and Germany), large contracts ('mega-outsourcing') have been agreed between a small number of multinational computer services suppliers and large client organizations, including central and local governments and large services and manufacturing firms. Many of these contracts are seen as having a number of problems, including excess fees, declining services, inability to adapt to changing business and technology needs, loss of power to monopoly suppliers and inability of the clients to manage the interface with the suppliers (Lacity and Hirschheim 1995; Lacity and Willcocks 2001; Willcocks and Fitzgerald 1994a). However, and despite the above problems, large-scale IT outsourcing is still continuing.

One of the problems in explaining the continuity of large-scale IT outsourcing is that existing studies apply theoretical approaches which offer limited explanatory power. For example, it is argued that firms externalize their IT activities because they can either save on costs/risks (the transaction cost perspective) or focus on their core competences (Lacity et al. 1994a). Little attention has been paid to wider changes in production systems. Indeed, while there have been a number of contributions examining the nature and impact of IT outsourcing (starting with Buck-Lew 1992; Lacity and Hirschheim 1993; Loh and Venkatraman 1992) and its implications for IT management (early examples include Grant 1992; Huber 1993; Quinn 1992), less attention has been paid to IT outsourcing in the context of broader organizational strategy and the implications for innovation and for the distribution of expertise in emergent organizational forms.

Because of the increasing importance of outsourcing of services, there is a need to understand why some services and product components are produced in-house and why others are contracted out. Much of the recent debate on specialization and deverticalization has revolved around the concept of modularity (Baldwin and Clark 1997; Langlois and Robertson 1995; Sanchez and Mahoney 1996). Virtually all the literature on modularity concerns manufactured goods. Examples of products with modular design include aircrafts (Woolsey 1994), software (Cusumano 1991; von Hippel 1994), personal computers (Langlois and Robertson 1992) and power tools (Utterback 1994). This chapter seeks to extend the research agenda by focusing on the lessons for modularity that can be drawn from the outsourcing of knowledge intensive business services (KIBS).[1]

KIBS are those services that support business processes and which require high levels of skills and knowledge, advanced technology (especially information and communication technologies) and strategic input (Miles 2001).[2] Examples include management consultancy services, computer and software services, legal and accountancy services, engineering services and research and development services. The rise of KIBS is, to a significant extent, the outcome of the increased technical and social division of labour within manufacturing production (Daniels and Moulaert 1991; Miozzo and Soete 2001). Given the rapid rate of technological change and the sophistication and variety of the services required, there has been a tendency to contract services from outside independent service producers or to set up subsidiary service firms. Thus, KIBS, which were historically 'internalized' in the large corporations have been 'externalized' since the 1980s. But a great deal of the growth of KIBS is more than simple 'externalization', as many 'outsourced' service functions are significantly different from those previously supplied in-house and respond to new needs by client firms (Miles 2003).

There is much evidence to suggest that client organizations benefit from external sourcing of KIBS. KIBS tend to be very IT-intensive, and are thus expected to play a desirable role in shaping economic growth through the diffusion of technology (Antonelli 1998; Katsoulacos and Tsounis 2000). They are also expected to promote investment in workers' skills (Peneder et al. 2000; Wolff 2002). Moreover, they form important intermediaries and nodes in innovation systems and may even complement the traditional 'knowledge infrastructure' of government labs, research organizations and universities (den Hertog 2000; Miles 2002).

However, there has been much less analysis of what happens at the interface of individual firms. One of the fundamental characteristics of KIBS is client participation in the production of the service (also called co-production). Because of the intangibility of services, uncertainty regarding the quality of services often requires close and continuous interaction between clients and suppliers (Miles 1993). Indeed, customer involvement in the provision of services has been referred to variously as interface, inter-action, co-production, 'servuction', socially regulated service relationship and service relationship (Gallouj and Weinstein 1997). This suggests that the capacity to interface effectively with KIBS will strongly shape the impacts of their use (Miles 2003; Miozzo and Miles 2002; Tomlinson 2001).

The literature on complex products and systems has pointed out an important limitation of the literature on modularity – that is, the adoption of a modular product architecture does not automatically lead to a modular organization pattern. Knowledge boundaries may be different from production boundaries defined by make-or-buy decisions, and know-ledge and organizational coordination demands interactive management, in many cases through 'system integrator' firms which 'know more than they do' (Brusoni and Prencipe 2001).

Drawing on empirical evidence on IT outsourcing in Germany and the UK, this chapter identifies an additional limitation of the modularity lit-erature that has potential applicability to a range of KIBS. Modularity is often presented as a design strategy that stimulates innovation. For example, Baldwin and Clark's (1997) study of the computer sector argues that through the adoption of modular designs, the computer sector dra-matically increased its innovation rate 'the fact that different companies . . . were working independently on modules enormously boosted the rate of innovation. By concentrating on a single module, each unit or company could push deeper into its working' (Baldwin and Clark 1997, p. 85).

Our research findings challenge the generalizability of this claim regard-ing a strong association of modularity and innovation when examining KIBS outsourcing. We argue that the intangibility of services, as deter-mined by particular features of services such as asymmetric information between client and supplier, product/service differentiation and dynamic scale economies (Sapir 1987), exacerbates the conflicts between clients and suppliers, which may present obstacles to innovation in a strategy of mod-ularization.

The chapter is organized as follows. The second section discusses previ-ous work on modularity and its limitations; the third section examines the peculiarities of IT outsourcing. The fourth section presents the research method and assesses the empirical evidence from 13 IT outsourcing con-tracts in Germany and the UK. Section 5 draws implications from the case of IT outsourcing for the potentially more general limitations to modular-ity strategies of KIBS.

VERTICAL DISINTEGRATION, MODULARITY AND INNOVATION

Vertical disintegration and specialization is perhaps the most significant contemporary organizational development of the corporation. IT is seen as a critical component both in driving organizational change and in enabling enterprise restructuring. To understand these developments, economists are reaching out for their dusty copies of the books of Adam Smith and Alfred Marshall. As argued by Krugman:

> the end of the corporation as we know it . . . a victim of information technology, which ended up deconstructing instead of reinforcing the corporation. . . . The millenial economy turns out to look more like Adam Smith's vision – or better yet, that of the Victorian economist Alfred Marshall – than the corporatist future predicted by generations of corporate pundits. (Krugman 1999)

Vertical disintegration and specialization can be seen at two levels. On the one hand, small entrepreneurial firms in traditional networks in Italy and Germany and the high-tech networks in Silicon Valley are able to exploit horizontal synergies of large organizations (Best 2001; Brusco 1982; Herrigel 1993; Piore and Sabel 1984; Sabel et al. 1989; Saxenian 1994). On the other hand, large firms in Japan decentralize their production to subcontractors, enjoying the scale of large firms and the flexibility of small production units (Aoki 1990; Best 1990). More recently, attention has turned to the vertical (and horizontal) disintegration of large US firms, which increasingly outsource and refocus their activities (Holmstrom and Roberts 1998; Prahalad and Hamel 1990; Zenger and Hesterly 1997). Shortening lifecycles, faster technology development and increasing product differentiation are seen as placing severe pressures on the capacity of the firms in all areas, with firms seeking to outsource 'non-core' functions. By tapping into production networks, leading firms are finding new ways to exert market power without the costs and risks of supporting a gigantic organization (Sturgeon 2002).

These changes in organization form and their impact on innovation have been analysed by a variety of authors. There is some agreement that internal organization is superior to arm's-length contracting for innovation. Different explanations have been offered for this. For Teece (1986), the key lies in complementary assets; for Silver (1984), the advantages are in informational terms. The danger of de-verticalization is that firms may shed competences that could become 'core' in the future. Indeed, Teece (1986) argues that appropriation of returns on core capabilities require access to complementary assets. Therefore, many non-core competences may be important too.

However, recent contributions in the innovation literature suggest that the large integrated firm may be less relevant to innovation in periods of rapid technological change. Langlois and Robertson (1992) suggest that large size and vertical integration are of little benefit in coordinating across the boundaries of the larger system, especially in the early stages of development, where experimentation is a much more important concern than coordination. When innovation is not systemic, it may proceed faster in a decentralized system because of its ability to utilize a more diverse set of information (Langlois and Robertson 1993, 1995). Also, the danger of de-verticalization can be averted by retaining 'absorptive capacity' (Nooteboom, Chapter 2 this volume).[3]

In complex products and production processes, there are strong technical interdependencies between what firms develop and make themselves, and what they require from their suppliers of machinery, components, software and materials (Grandstand et al. 1977). In addition to a focus on a number of distinctive technological competences, management in large firms needs to sustain a broader (if less deep) set of technological competences in order to coordinate continuous improvement and innovation in the corporate production system and supply chain. Furthermore, they must do this in order to explore and exploit new opportunities emerging from scientific and technological breakthroughs (ibid.). Indeed, the technologies and other capabilities required for certain innovations are now less frequently located within single firms and increasingly distributed across a range of firms and other knowledge-generating institutions. As a result of this increasingly complex division of labour in the generation of technology, innovations require the coordination of these existing capabilities and, possibly, the generation of new 'combinatorial capabilities' (Coombs and Metcalfe 2000).

The literature on 'modular production systems' recognizes this complex division of labour in innovation. It presents the characteristics and advantages of 'building a complex product or process from smaller subsystems that can be designed independently yet function together as a whole' (Baldwin and Clark 1997, p. 84). Modularity involves a clear architecture, clean interfaces, and a well-defined set of functional 'tests' of each module performance, where options in design are multiplied and decentralized (Baldwin and Clark 2000). Modularity is often presented as a design strategy conducive to innovation. Modularity is said to give full rein to the benefits of the division of labour by reducing the degree of interdependency among, and thus the costs of communicating across, the parts of the system (Langlois 2002). The advantages include that suppliers can improve capabilities narrowly and deeply (Langlois and Robertson 1995). A modular system is also said to open the technology to a much wider set of capabilities, and can benefit from external capabilities of the entire economy (Langlois 2003). Moreover, a modular system is said to facilitate the testing of many approaches simultaneously and allow for many entry points for new ideas (rapid trial-and-error-learning) (Nelson and Winter 1977) for autonomous (rather than for systemic) innovation (Chesbrough and Teece 2002). Modularity can generate 'economies of substitution' (Garud and Kumaraswamy 1995) or 'external economies of scope' (Langlois and Robertson 1995) which substitute for internal coordination. By designing modularly upgradable systems, firms can reduce product development time, leverage their past investments, and provide customers with continuity. Two important system attributes for realizing economies of substitution are 'modularity' (the ease with which system designers can substitute components) and 'upgradability' (the opportunity to work on an established technological platform) (Langlois and Robertson 1995).

Langlois (2003) has recently attempted to explain the origins of vertical disintegration (without rejecting Chandler's contribution). He argues that vertical disintegration ('the vanishing hand') is a further continuation of the Smithian process of division of labour in which Chandler's managerial revolution was a way-station (Langlois 2003). The 'vanishing hand' is driven not just by changes in coordination technology but also by changes in the extent of markets (see Adam Smith and also Young 1928):

> In many respects the structure of this new model looks more like that of the antebellum era than like that of the era of managerial capitalism. Production takes place in numerous distinct firms, whose outputs are coordinated through market exchange broadly understood. It is in this sense that the visible hand of management is disappearing. Unlike the antebellum structure, however, the new economy is a high-throughput system, with flows of work even more closely coordinated that in a classic Chandlerian hierarchy. (Langlois 2003, p. 373)

Langlois (2003) argues that decentralized production involves a broadening of capabilities and decoupling from specific products, carrying on from the Chandlerian firm. Just as did the high-throughput technologies of classical mass production, modular systems require and arise out of standardization albeit in a particular form. Langlois (2003) points out the difference with mass production:

> [U]nlike classical mass-production technologies, which standardize the products or processes themselves, modular systems standardize something more abstract: the rules of the game . . . By taking standardization to a more abstract level, modularity reduces the need for management and integration to buffer uncertainty. One way in which it does so is simply by reducing the amount of product standardization necessary to achieve high throughput. (Langlois 2003, pp. 374–5)

The concept of standardization of the rules of the game resonates in a variety of contributions in organization economics which suggest that vertical disintegration is made possible by IT and accompanying improvements in measurement and monitoring instruments. As argued by Zenger and Hesterly (1997) 'recent innovations and advances in measurement, monitoring, organizational design, and information technology have eased the selective infusion of market mechanisms into hierarchy and hierarchy into markets' (p. 210).

To this end, firms have developed new financial and nonfinancial performance measures: measures of quality, customer satisfaction, timeliness of delivery and innovativeness to measure sub-unit performance and mimic market mechanisms. In particular, the commonality of measures, which is critical to internal and external comparison, is also expanding, with consultants, industry associations and consortia responsible for their diffusion (Zenger and Hesterley 1997).

The literature on modularity suggests that participants can adhere to these rules and need not communicate the details of their production, producing a decentralized network and opening up the technology.[4] Nevertheless, it is not that the Chandlerian firm is giving way to pure 'modular production systems' and anonymous arm's-length markets. In many cases, the 'visible hand' has been socialized into standardization of 'interfaces' between production stages.[5] In an extreme case, standardized interfaces can turn a product into a pure 'modular production system or network' (Langlois and Robertson 1995), as is claimed by Sturgeon (2002) with regard to contract manufacturing in the US-led electronics industry. These networks imply heavy use of IT, as well as suppliers that provide widely applicable 'base processes' and widely accepted standards that enable the codifiable transfer of specification across the interfirm link:[6]

> These preconditions lead to generic (not product-specific) capacity at suppliers that has the potential to be shared by the industry as a whole and highly codified links between lead firms and suppliers that allows the system to attenuate the build-up of thick tacit linkages between stages in the value chain. Codified linkages allow the system to operate without excessive build-up of asset specificity and mutual dependence. (Sturgeon 2002, pp. 486–7)

It has been suggested that product architecture may have implications for organizational architecture (Henderson and Clark 1990; Sanchez and Mahoney 1996). Sanchez and Mahoney (1996) suggest that modular products facilitate the adoption of modular organization. In a similar vein, Arora and Gambardella (1994) argue that innovation increasingly relies on generalized and abstract knowledge, making knowledge more 'divisible'. They argue that this facilitates the separation of (tacit) knowledge (into 'modules') which can be the responsibility of different specialist organizations and that they can be reassembled at a later stage. Also, a more extensive division of labour leads to 'thicker' markets for information and knowledge-based services (encouraging de-verticalization because of reduced uncertainty in the reliance on outside providers and the incentive to develop the knowledge base of the firm in more general terms). Martin and Eisenhardt (2003) point out the advantages of organizational modularity (dedicated core initiative teams) and 'loose coupling' (simple coordination mechanisms) within multi-business organizations. Standardized committees facilitate the integration of activities with other groups, maintaining the specialization of dedicated teams, bringing focus and flexibility to adapt to changing internal capabilities and market needs.

However, the literature on innovation in complex products and systems (Davies and Brady 2000; Gann and Salter 2000; Hobday 2000; Prencipe 2000) underlines the limitation of these organizational implications of product modularity. Prencipe (1997) rejects the simple notions of core competences that recommend the outsourcing of production and, more importantly, the outsourcing of the development of components and subsystems, arguing that decisions based on economic factors alone may compromise the future technological competences of the firm. In his analysis of the technological competences of Rolls-Royce, he shows that this firm is vertically integrated regarding the components the performance of which influences the entire engine performance, such as the jet engine's inner core, and retains design and manufacturing capability and research. Moreover, although it contracts out the outer core, it maintains in-house understanding of the contracted technology to integrate it into the system.

Brusoni and Prencipe (2001) argue that modularity leads to increasing specialization and division of labour but that, contrary to some of the dominant accounts on modularity, this requires an intense effort of knowledge and organizational coordination: 'modular product architectures do not define clear "information structures" capable of coordinating actors via smooth arm's-length market relationships . . . product modularisation called for highly interactive organizational set-ups' (Brusoni and Prencipe 2001, p. 202).

This effort of knowledge and organizational coordination in 'loosely coupled' network structures is played by 'systems integrators', 'systems integrator firms outsource detailed design and manufacturing to specialized suppliers while maintaining in-house concept design and systems integration to coordinate the work (R&D, design, and manufacturing) of suppliers' (Brusoni et al. 2001, p. 613).

'Systems integrators' need in-house technological competences to absorb advances in knowledge in one part of the system or innovations emerging from specialized suppliers and to deal with imbalances in the performance of the components or in the architecture (Brusoni et al. 2001). Steinmueller (2003), however, underlines the difficulties involved in this coordination role. He addresses the problem of creating technical interface or compatibility standards (the collection of rules that enable components and subsystems to be assembled in large technical systems or complex products and systems) in supporting the division of labour. Indeed, he suggests that complex products involve a complicated mixture between interfaces that can be taken as a sufficient definition of the component or subsystem contribution to the entire system and interfaces that are 'incomplete' in defining the overall performance of the system. In the latter case, the definition of a technical compatibility interface is only a starting point for the design of the entire system.

This section reviewed the arguments in the innovation literature on how increased vertical disintegration, enabled by improvements in measurement and aided by the standardization of the rules of the game, can provide fertile ground for innovation (as opposed to the large vertically integrated firm). It has also surveyed the limitations of the organizational implications derived from product modularity identified by the literature on innovation in complex products and systems. This literature underlines the extensive efforts of knowledge and organizational coordination required in product modularity.

PECULIARITIES OF IT OUTSOURCING

IT outsourcing is the use of a third-party vendor to provide information products and services that were previously provided internally. IT outsourcing has grown in recent years to include multiple systems and significant transfer of assets, leases, and staff to a supplier who assumes profit and loss responsibility (Lacity and Hirschheim 1995). Computer services suppliers provide a number of services, including IT infrastructure (hardware and operating systems), applications development and support and helpdesk in exchange for a fee and according to a detailed contract.

The question of why firms are outsourcing the activities of their IT department at such an unprecedented rate when IT has never been more critical to business success has been widely explored. IT outsourcing is seen as key to business initiatives such as reengineering, knowledge management, the development of electronic channels of distribution and the development of digital business strategies (DiRomualdo and Gurbaxani 1998). Also, one of the most cited reasons for outsourcing is the need of client firms to refocus on core competences and the perception of IT as a cost burden (Lacity et al. 1994a). However, IT outsourcing also brings risks, including lack of competence of IT staff, loss of control, lack of organizational learning, loss of innovative capacity, and the lack of divisibility of IT: 'innovation needs slack resources, organic and fluid organizational processes, and experimental and intrapreneurial competences – all attributes that external sourcing does not guarantee' (Earl 1996, p. 30).

Standardization of the rules of the game is undoubtedly occurring – an attribute of considerable significance in descriptions of modular production systems. The outsourcing of IT services requires informed buyers, codification of processes, and contractual design that allow co-development and tailoring of services in well-governed inter-firm relations (Mahnke et al. 2003). Also, detailed contracts are generally developed in conjunction with tight performance measures designed to enable monitoring of services provision. This is in the form of service level agreements (SLAs). An SLA is an agreement that defines the parameters of the service including objectives, metrics, remedies and penalties for missing agreed levels of services provision. While different suppliers may have their own versions of this document, typical contents include: scope of the work; performance, tracking and reporting; problem management; compensation; customer duties and responsibilities; and warranties and remedies.

Nevertheless, there are two peculiarities associated with the outsourcing of IT that may prompt us to revise our understanding of the literature on modular strategies as applied to KIBS. First, IT is different from other organizational resources that have been successfully outsourced in the past. A number of features characterize IT: it evolves rapidly; its underlying economics changes rapidly; the switching costs to alternative technologies and IT suppliers are high; customers tend to be inexperienced with IT outsourcing; and it is IT management practices rather than economies of scale per se that lead to economic efficiency (Lacity et al. 1994b; Lacity and Hirschheim 1995; Lacity and Willcocks 1994). As such, arm's-length contracting may not be the more efficient practice for dealing with the provision and integration of fast-changing technology areas (Pavitt 2003).

Second, while growth of IT outsourcing and expansion of the supply base (with a growing number of specialist IT suppliers) may be easily interpreted as evidence of the separability of IT from internal production activities, numerous studies suggest this is not the case. While historically, IT was considered primarily as a support function (an administrative expense rather than a business investment), advances in IT mean that it now plays a critical role in strategy formulation and implementation (Venkatraman 1991). Indeed, the evidence from a growing literature underlines the fact that IT is important for the coordination of the firm. Jonscher (1994), for example, argues that in the modern firm it is increasingly difficult to distinguish between information and production technologies. Also, there is evidence to suggest that IT capabilities can be used to transform business structures and processes (Applegate 1994; Henderson and Venkatraman 1994), as well as to provide opportunities for increased connectivity, enabling new forms of inter-organizational relations and enhanced network productivity (Scott Morton 1991). As Mahnke (2001) argues, there is a need to recognize the embeddedness of IT and to reconsider its role within the firm in light of what he refers to as 'governance inseparability' and 'complementarity of capabilities' – notions that are neglected in conventional theories of the firm:

> Theories of the firm overlook [the fact] that a technologically separable interface between activities might not be available in codified form, and neglect learning dynamics that lead to strategic consequences in terms of capability development and adaptability in competitive environments of varying dynamics. (Mahnke 2001, p. 360)

The inseparability of IT from internal production activities means that even in situations of total outsourcing a minimum set of capabilities are often retained in-house by the client firm (the so-called 'residual IT organization') (Willcocks and Fitzgerald 1994b). As such, organizations may need to retain, change or develop different parts of their IT structures, capabilities and skills in order to maintain the linkages between IT provision and their business prerequisites. Recognition of the peculiarities of IT as an activity which is integral to the coordination of the firm means that the development of IT outsourcing poses questions regarding the boundaries, coordination and control of the modern firm. We explore these issues through presentation of our research findings in the next section.

IT OUTSOURCING IN GERMANY AND THE UK: EMPIRICAL EVIDENCE

Our research involved a study of IT outsourcing in two countries, Germany and the UK, selected since they represent the first and second largest markets for software and computer services respectively, in Europe (Ovum 2001). Moreover, in both countries, IT outsourcing has been the main driver of the growth in the IT market in recent years (Grimshaw and Miozzo, Chapter 6 this volume). We collected qualitative data through case studies of four computer services firms and a case survey of 13 client organizations which had outsourced IT to one of the four firms (see Table 4.1). The research also draws on secondary-source quantitative data for the IT outsourcing market in Germany and the UK, as well as organization- and contract-related documentation. The main advantages of this method include its sensitivity to the possibility of complex heterogeneous circumstances, its capacity to facilitate exploratory discovery and its suitability for analysing patterns across cases (Eisenhardt 1989; Yin 1981, 1994). The limitations of a case-survey approach are that the number of factors worthy of examination may be large relative to the number of cases selected and there is a risk that single factors are compared across cases in a way that oversimplifies the context of each case (Yin 1981).

We selected a sample of computer services firms that had signed large IT outsourcing contracts and reflected a mix of country of ownership. These included three firms in Germany (one US-owned, one UK-owned and one German-owned) and three firms in the UK (two US-owned and one UK-owned). For each firm in each country we sought access to senior project managers on the client and supplier sides of between two and three IT outsourcing contracts. We selected clients from a range of sectors of economic activity to match the structure of the client base in each country; public sector cases were therefore included in the UK and a greater proportion of manufacturing cases in Germany. We make no claims for the representativeness of the sample and thus have not sought to make generalizable conclusions.

The interviews were undertaken between September 2002 and September 2003. In the computer services firms we interviewed senior and executive managers with responsibilities for IT services operations to gain detailed information on firm strategy in the IT outsourcing sector. In both the client organization and the computer services firm, we interviewed the senior project manager responsible for managing the IT outsourcing contract. The intention was to explore the functioning of IT outsourcing from the perspectives of both organizations. Table A.4.1 lists the job positions of the 32 senior managers interviewed. All interviews were semi-structured and lasted between one and two hours. In agreement with the participants, we name the organizations investigated but preserve confidentiality in the presentation of qualitative data (numbering of cases does not correspond to the order in which they are presented in Table 4.1). Also, although there are differences in the initial inter-organizational arrangements adopted in Germany and the UK (Grimshaw and Miozzo, Chapter 6 this volume), we did not find significant differences according to country of ownership or country of operation with regard to questions of modularity.

We now assess the empirical evidence around four issues that follow from the above analysis of the lessons for modularity of outsourcing KIBS. First,

Table 4.1 IT outsourcing cases in Germany and the UK

IT supplier	Client organization	Type of outsourcing	Value of contract(s)[a]	Duration in years of contract (s) (start date)	Staff transfer
IBM (DE)	Hapag Lloyd	IT infrastructure[b] (including desktops), help desk, service management and network system management	€100m €80m	5 (1998) 5 (2002)	20–30
IBM (DE)	Deutsche Bank	IT infrastructure (including desktops), help desk, service management and network system management	€35m €2500m	5 (1999) 10 (2003)	900
IBM (DE)	Continental[c]	Data centre management	€75m	10 (1995)	280
Logica (DE)	Becks (Interbrew)	Applications support and development	€54m	5 (2000)	28
T-Systems (DE)	Henkel	IT infrastructure (including desktops) and service management	€100m €35m p.a.	7 (1996) 7 (2000)	55
T-Systems (DE)	Deutsche Telekom[c]	Desktops, network system management and applications support and development	n.a.[d]	(2001)	Approx. 12 000
IBM (UK)	AstraZeneca[c]	IT infrastructure (including desktops and disaster recovery) and service management	£1150m	7 (2001)	420
IBM (UK)	Cable&Wireless	IT infrastructure (including desktops), helpdesk , service	£3500m	10 (1998)	1 600

94

EDS (UK)	Dept of Work & Pensions[c]	IT infrastructure (including desktops), network system management, and applications support and development	> £ 2 000m[e]	10 (2000)	3 000
EDS (UK)	Inland Revenue	IT infrastructure, service management, and applications support and development	£ 1600m	10 (1994)	2 300
EDS[f] (UK)	Rolls-Royce	IT infrastructure (including desktops), network system management, and applications support and development	£ 1400m	10 (1996–97)	1190
Logica (UK)	Britannia Airways	Applications support and development	£ 27m	7 (2000)	30
Logica (UK)	British Petroleum Chemicals	Applications support and development	£ 27m (10 years)	5 (1995) 5 (2000)	20

Notes:
a Initial nominal value of contract and initial nominal value of any contract renewal (excludes additional work).
b IT infrastructure includes hardware, operating systems and databases.
c We were unable to arrange an interview with a representative from the client organization.
d We were not informed of the total value but that the arrangement involves more than 4500 contracts, with the largest valued at €400million p.a. and the smallest at €10000 p.a.
e This cost estimate is based on public information, which states that prior to signing the contract the Department of Work and Pensions spent over £200m annually on IT services (the interviewee confirmed that that figure represented the annual cost after the contract). Also, while EDS is the lead supplier, it operates on this contract jointly with IBM.
f This contract involves a merger of three separate contracts, two negotiated in 1996 and one in 1997.

Source: The Financial Times (various issues); management interviews; company documentation.

we examine the standardization of the rules of the game and performance gains. Second, we illustrate the way in which, because of the inseparability of information and production technologies, IT outsourcing is accompanied by wider changes in production technology. Third, we examine the particular forms of organizational and knowledge coordination and control in IT outsourcing: the retained IT organization and staff transfer. Finally, we suggest that the intangibility of IT outsourcing services exacerbates tensions arising from conflicts between client and supplier organizations that may hinder innovation.

Successful Modularization Design? Standardization of the Rules of the Game and Performance Gains

In all 13 case studies there is evidence of standardization of the 'rules of the game' and gains from specialization in the provision of IT services. First, all cases had relatively standard contracts, typically comprising several service level agreements (SLAs) within a Master SLA. Third parties were often hired to help draft, evaluate and negotiate the contract; expert consultants for the sector, such as Gartner, thus diffused similar procedures and methods. The agreements were very extensive, reflecting the high value (more than €1 billion in six cases) and long duration (between 3 and 10 years) of contracts:

> The contract is the key. It is the cornerstone of the whole business deal. (Case 1, IT supplier, UK)

> [The contract] is regarding programme management, how we will jointly do projects, it's regarding customer service centres, service management, asset management, data centre infrastructure management, system operations and control, IT security, disaster recovery, subcontractors, policy standards. . . . So it is a bridge between [the client's] processes and [the supplier's] processes. . . . The [client's] road map is discussed here, product catalogue is discussed here, how we manage utilities, how we manage service levels, programme management. . . . It's 300 pages or so. (Case 13, Client, Germany)

Second, contracts typically involved provisions for fixed and flexible tariffs, allowing for negotiated additions, changes or deletions from the Master SLA. A fixed (and usually cost-reducing) tariff is applied to specified baseline services and may also specify a certain number of daily rates for consultancy services from the IT supplier. If an IT supplier fails to meet key performance indicators specified in the SLA, then typically it must pay a cash penalty. If additional services are requested, then they are generally priced by the IT supplier. Thus, in all 13 cases, client organizations applied relatively standard metrics to measure and monitor IT service performance, the number of technical problems and the time it takes to fix them. For example, the standard practice of 'function points' (originally developed by IBM in the 1970s) was used to measure software development effort.[7] Several other metrics systems were also built into the contracts:

> We also use a tool called Cocomo, which came out of South Carolina University. It assesses 30 different influences on software productivity. . . . We were able to use that to validate the increase of manpower that [the IT supplier] was asking for. (Case 2, Client, UK)

> [Targets set for the supplier include] setting up some new emerging technologies, showing that they are working with initiative, . . . and that they are implementing new innovative technologies. We have an innovation steering committee looking at these issues. (Case 12, Client, Germany)

Third, all client organizations benchmarked prices and developments in new technologies through use of international consultancies such as Gartner, Compass, or Capital Procurement; in some cases, they also utilized information drawn from open-book accounting with the supplier firm. The above – standard contracts, performance measures and their diffusion through international consultants – point to evidence of standardization of the 'rules of the game' between firms.

Data from all 13 organizations, also support arguments for the direct performance improvements following IT outsourcing. A good part of this was due to cost savings arising from economies of scale and production efficiencies. Strategies employed by the IT suppliers included consolidating data centres (e.g. from 12 to 4 in Case 2, UK and from 5 to 2 in Case 5, UK), rationalizing IT applications around fewer servers, achieving better prices with suppliers of hardware and on licence agreements, and reducing staff costs.

But direct improvements in IT were also crucial to client managers' perceptions of improved performance. In some cases new technologies were wholly designed and developed by the supplier; in others they were adapted from an off-the-shelf product, and in some the supplier implemented an application developed by the client. For example, in Case 2, UK, the supplier applied its expertise in rolling out, networking and supporting 70 000 new desktops in less than 12 months with no disruptions to normal work. In another case, the client airline company benefited from new Personal Digital Assitants (PDAs) issued to all flight crew, enabling them to connect to e-mail and download the 'flight pack' from the main data warehouse; moreover, a new seat selection system allowed the company to charge customers £10 each for the option of selecting their seat on a flight. Also, in the case of the liner shipping company, 'a big step forward' came from implementing the standard platform – the PC LAN infrastructure – around its worldwide operations. This was made possible by linking up with a multinational IT supplier with global presence. The standard platform meant upgrading IT was much faster; we were told that software upgrading would take 24 hours, compared to anything up to two weeks prior to the outsourcing arrangement. There were also several examples of bespoke innovations, such as one of the chemicals firms that benefited from the development of a vendor inventory management system designed especially by the IT supplier. This involved installing monitoring devices on the walls of tanks of chemicals, which would trigger automatic orders once the level dropped below a certain minimum.

Taken together, evidence of standardization of the rules for inter-organizational contracting, coupled with direct improvements in IT systems, may suggest a successful design strategy of modularization in IT outsourcing, including the benefits of the division of labour, with performance gains through specialization (section 2). However, further exploration of the data suggests reason for caution. Because the externalized IT functions remain embedded within the business processes of the client organization, there is potential for significant indirect transformation in the wider production system. The inseparability of IT from production means that suppliers are not 'turnkey' – where clients can easily substitute components – raising the risks of 'lock-in'. Moreover, standardization of the rules of the game comes at a cost. The 'governance inseparability' (section 3) between client and supplier demands considerable investment in interpersonal and administrative relations between firms, as well as in systems of monitoring and measuring. In addition, on the one hand, the client must retain some minimum capability to manage the organizational coordination of IT with its internal business strategies, and, on the other, the supplier requires staff transfer to provide client-specific skills for knowledge coordination. The intangibility of IT thus generates tensions where the potentially conflicting business strategies of client and supplier become transparent. This may hinder innovation. We proceed to examine each of these issues in turn.

Inseparability of Information and Production Technologies: IT Outsourcing Accompanied by Wider Changes in Production Technology

Our case studies attest to the ubiquitous penetration of IT in all business functions and the inseparability of information from production technology. In eight of the 13 cases, IT outsourcing was accompanied by a significant change in the wider production technologies of the client organization (Table 4.2):

> IT itself is a central nervous system for the business processes. And you are outsourcing maybe the heart for running this central nervous system. And that meant for the company, also for the employees, a real, very deep cut, and a cultural change, in how to run the business. (Case 12, Client, Germany)

Table 4.2   Cases where IT outsourcing was accompanied by significant change in the client's production system

Case 2, UK	New IT systems facilitated speed-up of product lead-time, but only through pressures on government ministers with responsibility for this public sector client to reach decisions more quickly. The aim was to introduce a more transparent tradeoff between costs and consensus decision making in government.
Case 3, UK	New flight and flight crew management systems enabled reductions of average size crew per aircraft. New aircraft maintenance schedules enabled rationalization of engineers.
Case 6, UK	IT outsourcing was explicitly associated with a 'business transformation programme', involving change in product development, manufacturing processes, supply-chain management, repair and overhaul (inventory management), workers' employment conditions and project management. Results included reduced product development cycle, reduced assembly time, factory closures, reduced number of suppliers, and new 'output-based' manufacturing plants.
Case 7, UK	Common data warehouse enabled client to assess performance across its worldwide operations more systematically. Vendor inventory management system speeded up throughput and increased revenue. Centralized European sales office for customer relations.
Case 8, Germany	Expanded (and diversified) retail sales through e-business. Additional applications to SAP (e.g. wireless stations on fork lift trucks) speeded up throughput.
Case 9, Germany	IT systems facilitated measurement and scheduling of unit volumes of shipping cargo; centralized system provided competitive advantage in terms of worldwide coordination.
Case 12, Germany	IT improved lead time of research projects and the 'time to market' of new products.
Case 13, Germany	IT contract involved 'run the bank' and 'change the bank' activities. Developed new IT-based service products.

Perhaps the most dramatic case is that of the aero-engine producer, in which IT outsourcing was accompanied by a complete business transformation. The supplier assumed responsibility for all technology infrastructure, networks, systems, and applications of the client. This was done with a business consultant's support in change management (business process re-engineering). According to joint client/IT supplier documentation, business transformation included initiatives in product development (product development lead time reduced from 48 to 34 months, 16 per cent increase in engineering productivity), lean manufacturing (cut engine build times by 50 per cent, 30 per cent increase in stock turns and 30 per cent reduction of work in progress), supply chain management (ordering of multiple parts under one part number to enable close relations to first-tier suppliers), strategic sourcing (long-term machine tool deals with preferred suppliers) and repair and overhaul (improved turn-around time and reduced inventory pools by a factor of three). Significantly, the technological improvements in the engines forced a change in the overall business model, from the sale of engines to the leasing of engines – sale of 'power by the hour' (Case 6, Client, UK).

> [The client] was running at about 2 per cent of its overall turnover invested in IT. So the discussion went along the lines of, 'Well OK, if you outsource your IT functions, you will gain some efficiencies and some cost savings, but it's only on 2 per cent of the overall turnover. So it's worth having. But how will that get you ahead of [competitors]? . . . Instead of just looking at IT as a separate thing, could we not join the business strategy and make some improvements in the way the operation works, supported by new IT, not just running IT as it was'? (Case 6, IT supplier, UK)

Similarly, in the case involving a liner shipping company, IT was considered crucial in transforming its business model by facilitating improved measurement of business performance:

> We measure our business in 'TUs'. TU means 'twenty feet equivalent unit'. . . Every year we have an increase of about 10–12 per cent of these TUs and we do that with the same staff every year, so the productivity increase is dramatic and for that reason we need good accuracy, IT systems and also an excellent IT operation. (Case 9, Client, Germany)

New IT systems increased the capacity to handle bookings and also improved coordination with other shipping companies, enabling sharing of cargo space. The worldwide, centralized system implemented by the multinational IT supplier was considered a major competitive advantage by the client organization.

In the case involving a charter airline, prior to the IT outsourcing arrangement, the ground handling staff at the airport would telex the client with information regarding flights landing. All flight details were tracked on wipeboards on the walls of offices. New IT systems transformed this activity so that all operations were subsequently undertaken in a small office with six double-screen workstations. Further applications included systems for managing cabin crew leave and scheduling of aircraft engineering maintenance. Combined, the client benefited from important cost savings, including a reduction in the number of crew members deployed per aircraft from 3.7 to 3.2, generating 'mega, mega savings' (Case 3, Client, UK).

In a final example, the IT supplier sought to overcome the problems of ERP (Enterprise Resource Planning), which had handled data in a way that was not so useful for the management of the client organization. It developed a common data warehouse that enhanced the client's ability to manage the business.

> [The IT supplier] developed this common data warehouse using various techniques, various pieces of software, which gave [the client] a much improved way of viewing the business globally . . . to see the whole business or to cut it up into pieces, or by salesman, or by area . . . We could see what it was costing us right through the supply chain from manufacture through to selling. . . . You could begin to see how much it was costing us to service a particular customer which then gave you a good idea about whether we were making good margins with that customer. (Case 7, Client, UK)

Overall, therefore, the ubiquity of IT in the organization and the inseparability of information and production technology meant that IT outsourcing was often accompanied by wider transformations in the production systems of the outsourcing firm. Our evidence suggests that IT outsourcing had an impact on a variety of areas associated with the client organization's services delivery or management processes. A consequence of this is that there was great demand for organizational and knowledge coordination. We address this issue in the next sub-section.

Forms of Organizational and Knowledge Coordination: Staff Transfer and Retained IT Organization

While standardization of contracting rules facilitated the outsourcing arrangement, the pace of change of IT, coupled with its inseparability from change in wider production technologies generated incentives in all cases for the client organization to institutionalize new forms of coordination and control. As such, despite the emergence of standardized 'rules of the game' there was a demand for a great deal of investment in time and resources for organizational and knowledge coordination. A first dimension to this involved active management at both the contract and relationship levels. Client and supplier managers stressed the need for frequent discussion, negotiation and renegotiation to reconcile objectives between the client and supplier and to institutionalize a process for managing conflict. IT outsourcing thus required much management input, not little, as expected in the dominant accounts of decentralized systems. Indeed, intense communication of information is not regarded as contradictory with modularity (as shown by Langlois' comments on modularity in the automobile industry), 'it would be a mistake, I think, to read the essence of the new collaboration in automobiles as arising out of a demodularization in which encapsulation has been eliminated in favour of intense communication. Quite the opposite is the case' (Langlois, 2002, p. 34).

Our data suggest that management input and intense communication was especially important in Germany. In Case 9, there were daily (service status report, involving a team of five from the client), weekly (service status report and action items), monthly (review of service and SLA fulfilment) and twice yearly meetings (steering committee). In Case 12, there were weekly (operational meetings), monthly (market developments, costs and pricing discussions) and yearly meetings (discussion of general service fee for systems administration). And in Case 13 there were weekly (sometimes daily) conferences, boards to discuss the weekly issues and a senior management board every three to four weeks. Frequent interpersonal communication allowed managers to practise active conflict resolution. Indeed, this may be evidence of the relevance of the principles of modularity since 'standard' committees may allow the integration of activities with other business groups.

A second, and vital, dimension to new organizational and knowledge coordination strategies utilized by the client organization is the in-house retention of capabilities and skills – the so-called retained IT organization, or interface. Retained skills included not only technical (operational) know-how and contract/service management capability, but also strategy and leadership skills. There were different strategies across the 13 cases (Table 4.3). In Case 6, no staff were retained in-house, ostensibly as a sign of interdependent and strong trusting relations between the two firms. By contrast, the client organizations in Cases 1 and 4 not only had a retained organization, but increased its size during the initial stages of the contract in order to handle the unexpectedly large volume of information.

In its positive role, we found evidence in some cases where the retained IT organization was seen not only as providing for a bureaucratic check on

Table 4.3   The 'retained IT organization' in client organizations

	Initial number	Comments
Case 1, UK	6	Increased to 75 employees within first 18 months; subsequently reduced to 8.
Case 2, UK	30	Takes role of 'intelligent customer'.
Case 3, UK	17	Strong internal team to manage all IT outsourcing contracts. Six of the 17 IT people retained in-house are senior managers.
Case 4, UK	n.a.	Increased in number within the first 12 months; high use of freelancers.
Case 5, UK	n.a.	In-house IT group and 'external supply group' to monitor the contract.
Case 6, UK	None	Reflection of high trust relationship.
Case 7, UK	n.a.	A 'small group' retained to manage the contract; also use external consultants.
Case 8, Germany	4	–
Case 9, Germany	6	Team of specialists managed by an externally recruited outsourcing specialist.
Case 10, Germany	70	Manage the contract and encourage innovation transfer.
Case 11, Germany	n.a.	–
Case 12, Germany	55	Team of specialists retained.
Case 13, Germany	30–40	Mix of business managers, IT managers and customer relations managers.

the terms and conditions of the contract (thus reducing the risk of opportunistic behaviour by the supplier), but also as essential for developing different parts of the IT structure and coordinating the link between IT and business strategies. The following quotes are illustrative:

It gives them somebody on their side who speaks the same jargon as the IT people and who then explains it back to them [the client]. . . . It also gives them somebody who they trust. (Case 1, IT supplier, UK)

We definitely do not want [the supplier] defining the strategy concerning the infrastructure for us. . . . We need competence on the contract management on our side to have a balance. (Case 9, Client, Germany)

However, in both countries, it could also act as a source of conflict, largely because it was often constituted not only by IT staff in the client firm (who were not transferred to the IT supplier), but also by external consultants who may have had several ulterior motives. In Germany, managers from IT suppliers referred to the three-way relationship as the 'bitter triangle' (Case 13, Germany), and the retained IT organization as 'the freaks' (Case 9, Germany) and the 'brick wall' (Case 1, UK). In Case 7, UK, the IT supplier complained that freelance consultants working in the client's interface often cherry-picked new projects for applications development. In Case 1, UK, the IT interface expanded spectacularly from 6 staff to 75 within the first 18 months of the contract. The main reason was a perceived need to establish greater control over contract verification procedures, involving evaluation of long and detailed monthly performance reports from the supplier. The rapid expansion of monitoring work forced the client to buy in expertise from outside, including IT consultants subcontracted from firms that were in competition with the IT supplier. From the IT supplier's perspective, they not only had to deal with an increasingly bureaucratized interface, but also faced monitoring by a rival firm. This problematic relation imposed a high price:

> With the 75 in place we were paying more penalties because there was a hold-up of things. Every change that we raised – be it an emergency or a normal change – was waiting on these guys to . . . approve it. . . . Because of this, [the interface] gained a momentum and autonomy of its own. They were making judgments on behalf of the business as to what was accepted or rejected. (Case 1, IT supplier, UK)

A third dimension to new coordination strategies provided client managers with a degree of control over IT skills and capabilities – the practice of staff transfer. In Germany and the UK, staff transfer underpins IT outsourcing. IT suppliers bid for IT contracts on the assumption that incumbent workers will transfer to the IT supplier that wins the contract. The number of transferees among the 13 cases ranged from 20 to around 3000 (Table 4.1). There was a consensus among IT suppliers that the transferred employees were critical to bringing the required industry- and firm-specific expertise in IT systems, as well as knowledge of the client's business processes:

> We need these people [transferred staff] because they have a lot of experience in the [client] processes. They are good guys, . . . The most important thing for me, they have a very good customer relationship. (Case 12, IT supplier, Germany)

> [The contract cannot operate without staff transfer] because it needs to have the skill from the data centre, all the jobs [which control how applications interact]

and all the applications, in which order they are to be done, and so on . . . the whole operation . . .. (Case 13, IT supplier, Germany)

This attests to the need for knowledge coordination in inter-organizational arrangements in a decentralized system (given the difficulties of codifying this knowledge). While staff transfer meant that the client (as the ex-employer) still benefited from the accumulated tacit knowledge, and its own past investments in skills, such benefits diminished with the duration of the contract. First, many transferred staff sought to broaden their experience of IT and moved into other areas (and countries) of the IT supplier's business operations. Second, the IT supplier often exercised its prerogative (as the new employer) to 'creamskim' the more able staff in order to maximize added value across the range of business activities. Third, transferred staff sometimes resisted the change of employer and sought to undermine efforts by the supplier to improve performance. And finally, there was some evidence that investment in skill development followed a trajectory of peaks and troughs as IT suppliers faced the risk of losing contracts to competing suppliers, thus diminishing their ability to recoup the costs of investing in skills.

As suggested by the literature on complex products and systems, product/services modularity may require very interactive organizational set-ups. Given the inseparable nature of IT from production technology, there may be new dimensions to client organization strategies to control and coordinate knowledge and business activities in line with the new IT outsourcing arrangement. In each IT outsourcing contract, these dimensions – the meetings and committees, the retained IT organization and staff transfer – demanded considerable investment and provides additional support to the claim of the literature on complex products and systems regarding the limitations to the organizational implications of product modularity. The next sub-section explores another particular feature of IT outsourcing, the intangibility of the service provided, and its effect on innovation.

IT Outsourcing and Intangibility: A Case of Blocked Innovation?

The intangibility of IT services (determined by asymmetric information between client and supplier, product/service differentiation and dynamic scale economies) leads to uncertainty regarding the quality of services provided and exacerbates many of the tensions arising from conflicts between client and supplier. Modularity is often presented as a design strategy which is innovation-friendly (second section). However, evidence from the 13 cases suggests that innovation, as organized through IT outsourcing, may be blocked due to the conflict between the cost-based nature of the contract and the flexibility and freedom required for innovation. Table 4.4 lists examples of tensions arising from this conflict.

A first example of tension concerns the client's goal of reducing its cost base and the supplier's aim of expanding revenue and profit margins by

Table 4.4   Examples of conflicts between client and supplier organizations

Case 1, UK	Cost reductions from rationalizing number of servers not passed on to client; conflicts over knowledge transfer because the supplier contracted with a second client that competed against the first; 18-month renegotiation of contract, ultimately led to decision to return all IT in-house.
Case 2, UK	Client complaints that supplier was insufficiently innovative and that it recruited staff with inadequate qualifications and experience to save on labour costs (registered in quarterly performance reviews); all upgrading required additional payments by client; conflict over who paid for training during final period of contract.
Case 4, UK	Client concerned about international downsizing strategy of IT supplier, leading to redundancies of key IT staff.
Case 6, UK	Tension between client's cost reduction strategy and supplier's revenue expansion strategy; tension around shift to off-the-shelf IT services.
Case 7, UK	Client initially reluctant to invite supplier to IT strategy meetings due to suspicion of opportunistic profiteering; supplier frustrated by client's purchase of SAP from a competing supplier, making it more difficult to bid for development work; contractual misunderstandings regarding costings for software development.
Case 8, Germany	The initial marrying of business objectives through a local joint venture complicated by buy-out by IT supplier. The client managed the purchase of SAP and other software licences, frustrating the supplier's capacity to control the IT strategy.
Case 9, Germany	Client devoted resources to pressure supplier for innovation, and also had to ensure the supplier committed high-quality skills to the contract; strong cost-led nature of the contract impeded supplier's ability to innovate; international restructuring within IT supplier shifted some activities to Hungary, generating coordination problems and a decline in service quality.

Table 4.4   (continued)

Case 11, Germany	Tension between cost reduction through standardization of IT systems versus client attempts to retain control of its firm-specific technologies.
Case 12, Germany	Supplier rationalized data centre to serve more than one client, but the lack of exclusivity led to delays in managing services delivery; client's strong focus on costs impeded supplier's ability to increase revenue by selling additional IT services; client pressure to develop offshore contracts, but supplier's concern over global coordination costs; client not satisfied that supplier had sufficient global expertise.
Case 13, Germany	Conflict over switch to standardized hardware and also over which organization ought to finance the retained customized technologies.

selling additional IT applications, or upgrading systems. In two German cases (Cases 9 and 12), managers in the IT suppliers argued that the strong cost focus of the agreed contract impeded the introduction of new technologies:

> If you want to bring new technology into an environment it will cost you some money which somebody has to pay. If the contract is not calculated that comfortably that it brings that money, you can not do that, you have to run it on the old machines until they are paid off. (Case 9, IT supplier, Germany)

In two UK cases (Cases 1 and 7), there was apparent evidence of opportunism by the IT supplier as it sought to obscure the true cost savings achieved from rationalization of IT services, fulfilling its goal of profit maximization, but at the risk of harming its reputation as a trustworthy partner:

> From the point of view of doing things without the customer seeing, you can sometimes move things from one box to another. As far as they are concerned, you are still running six UNIX boxes but in effect we are running all six applications off one UNIX box, because we just brought a bigger one in. Running all applications from one box gives us two major advantages: (a) they run faster because they are on a bigger box . . . , and (b) you then have spare box[es]. (Case 1, IT supplier, UK)

A second tension concerns expectations of innovation. Interview and documentary evidence in three case studies (Cases 2, 4 and 9) strongly suggest that the supplier had not fulfilled the client's expectations about the quantity and quality of innovations. In some cases this reflected the supplier's weak capability to bring new innovations to the client's attention, while in others it resulted from a deterioration in the quality of personnel allocated to the contract by the IT supplier:

> Before outsourcing, . . . we thought that [the IT supplier] would always want innovations and that we would have a problem to follow. . . . After outsourcing it was totally different . . . [The supplier] is very passive . . . It's a lot of effort and power from our side to get innovation, to push them. (Case 9, Client, Germany)

Wider restructuring by the IT supplier – to cut costs and to streamline international operations – often conflicted with the mutually nurtured, 'joint' goals of the local management teams with responsibility for the outsourcing agreement. In Case 4, a major downsizing exercise by the IT supplier in 2002 caused consternation among managers from the client organization (IT supplier, UK). And in Case 9, the relocation of some IT services from Germany to Hungary decoupled the strong interpersonal ties, which had underpinned high-quality services provision (IT supplier, Germany).

Aside from the conflict between cost and profit objectives, our evidence reveals two further difficulties in designing contracts for innovation. The first is that innovations may be hindered by a conflict between the client's desire to retain, and develop, customized technologies, and the supplier's goal of standardizing technologies as a means to spread cost savings over a range of contracts with different client organizations. In three cases (Cases 6, 11 and 13), there was clear evidence of IT suppliers who sought to develop the contract through providing standardized products and processes, but against the client's wishes:

> [The client] was very much in the vein of writing bespoke applications. . . . They regard themselves as different from everybody else. One of the key things about this [IT outsourcing contract] was proving that they are not different. . . . That did cause some conflict. (Case 6, IT supplier, UK)

> If [the supplier] says standardisation [and] the customer says no [to] this process, not at this place . . . [then it] causes more problems. We call the customer and discuss whether to make a decision to pay more or to take the [off-the-shelf] process. (Case 13, IT supplier, Germany)

The second difficulty of contracting for innovation was caused by the absence of clauses, in some UK cases, pertaining to the costs of training IT personnel. This was most evident in Case 2, where the ten-year contract was drawing to an end and the future supplier had not yet been agreed. There had been disagreement between the two partners regarding who was responsible for financing skill development during the final three years of the contract, since many of the 3000 staff committed to the contract would transfer to the new supplier with the award of a new contract. The IT supplier refused to pay because it couldn't guarantee a return on its investment; client managers were told that the CEO of the IT supplier 'would not sign it off' (Case 2, Client, UK). Ultimately, the client organization agreed to finance the skill development programme. Paradoxically, because improved skills among employees of the IT supplier contributed to subsequent productivity improvements, the client also had to make additional bonus payments to the supplier because performance exceeded that stated in the SLA.

A third tension between client and supplier objectives concerned the difficulties of managing the joint relationship alongside other (often overlapping) contractual arrangements with other suppliers, and other clients. In four cases (Cases 3, 7, 8 and 12), client organizations had adopted multi-supplier contracts, partly to reduce the risk of 'lock-in' to a single, powerful IT supplier.[8] The effect was to increase pressures on suppliers to reduce costs. In those cases without multi-supplier contracts, client managers were more likely to voice concerns over the dangers of 'lock-in' – in particular the risk that their innovation strategy would be tied to the goals of a single IT firm. Also, the strong penetration of IT (and therefore the danger of lock-in with IT suppliers) into clients' business processes forced some client organizations to invest considerable resources in 'visiting and buttering up potential bidders', to persuade them of the merits of competing for the next tender (Case 2, Client, UK).

Multi-client contracts held by the IT supplier firm also caused problems. In Case 1, UK, the supplier won a new contract with a firm in direct competition with its existing client, leading to fears among client managers of knowledge transfer and replication of IT systems that had delivered it a competitive edge. And in Case 12, Germany, the client had experienced a deterioration in the number of services warranting 'high priority', due to the supplier servicing multiple clients from the same data centre.

Overall, therefore, although IT outsourcing was associated with enhanced performance (through economies of scale, production efficiencies, introduction of new technologies), the sustainability of innovation was brought into question by conflicts of business objectives between the client and IT supplier. These tensions were exacerbated by intangibility and uncertainty regarding the quality of the services. In part, these conflicts were fuelled by the form of the contracting arrangement, which required careful monitoring by the client of the cost-based targets; attention on a short-term bottom-line hindered joint innovations and upgrading of IT systems and IT-enabled products and processes.

IMPLICATIONS OF IT OUTSOURCING FOR THE MODULARITY OF KNOWLEDGE INTENSIVE BUSINESS SERVICES

Baldwin and Clark (1997) identified the relevance of modularity for services. They argue that a range of services are also being modularized, especially in the financial services industry. In their account, it is intangibility that facilitates modularization, 'financial services are purely intangible, having no hard surfaces, no difficult shapes, no electrical pins or wires, and no complex computer code. Because the science of finance is sophisticated and highly developed, these services are relatively easy to define, analyse and split apart' (Baldwin and Clark 1997, p. 88).

However, our research calls into question the generalizability of such claims. The inseparability of IT from production technology makes it difficult to 'split apart' computer services from the rest of the operations and to provide them by an independent supplier through arm's-length contracting without an intense effort of organizational and knowledge coordination.

Organizations may be attempting to modularize certain services for many of the same reasons that they aim to modularize manufactured products. The emphasis on specialization is widely adduced as a reason for the growth in KIBS – that specialized firms can supply services more efficiently and achieve greater economies of scale than in-house services. Indeed, our research data show that outsourcing of IT, as an example of design modularity of KIBS, can enable economies of scale, production efficiencies and the introduction of new technologies. However, recourse to external KIBS involved not just a simple substitution of internal services but instead a rather more complex process of knowledge transfer that required reciprocal learning and interaction (Miles 2003). As such, outsourcing of IT raises a number of questions about the structure, coordination and control of the modern enterprise.

In eight of the 13 cases, inseparability of production and information technologies meant that IT outsourcing was accompanied by wider transformation in the client's business. Importantly, these changes were not simply a consequence of more sophisticated IT systems, but also reflected the fact that new IT systems were delivered through an outsourcing agreement. As such, the focus of change was towards improved measurement and monitoring of a range of areas of business performance, in line with the metrics developed in the course of developing and running the IT outsourcing contracts. Indeed, management of IT outsourcing relations means that the client is therefore concerned not only with relations with the external supplier, but also with knowledge and organizational coordination and control (Brusoni and Prencipe 2001) of its own operations. Knowledge and organizational coordination strategies in IT outsourcing involved two particular practices: an in-house 'retained organization' to coordinate the link between the IT and business strategies of the client and the transfer of employees with industry- and firm-specific expertise.

The modularity literature argues that what makes decentralization economically efficient is the possibility of a standard interface that allows the modules to coordinate with one another without communicating large volumes of information (Baldwin and Clark 2000). In certain types of manufacturing, advanced technological knowledge about component interactions can be used to fully specify and standardize the component interfaces that constitute a modular product architecture, creating a nearly independent system of loosely coupled components that can be produced and bought from separate organizations. However, this may be more challenging in services because of the need for close and continuous interaction between client and supplier. The case study data attest to a wide and growing supply base of computer services providers – with suppliers deepening their competence and increasing their scale of operations – and the development of standard contracts and accompanying improvements in measurement and monitoring instruments. However, the need for 300-page contracts and the retention of dozens of specialist managers in the client organization meant that the interface between client and supplier organizations required extensive attention and management effort, mainly due to the inseparability of IT and production technologies and the intangibility of services. This involved considerable build-up of 'asset specificity' and 'mutual dependence' (Sturgeon 2002), and thus less flexible substitutability of components (in effect, services modularity but less 'organizational' modularity) than suggested in the literature on modular systems in manufacturing (second section).

Evidence from our case studies demonstrates that the sustainability of innovation is brought into question by intangibility of services, which exacerbated conflicts between the client and supplier organizations. In some cases, the cost focus of the contract impeded the introduction of new technologies. In others, there were tensions between the client's desire to retain customized technology and supplier's aim to standardize technologies across a range of contracts and clients. Also, in some cases, client expectations of superior innovation development were not met due to a deterioration in the quality of IT workers allocated by the supplier to the client's business. These conflicts may have been fuelled by the client's main focus on careful monitoring of the costs in the contract. This goes contrary to the general claims in the literature which present modular design strategies as very conducive to innovation.

CONCLUSION

Deverticalization is perhaps the most important organizational development since the 1990s. While a number of studies have argued that modularity may be conducive to innovation, our evidence from 13 large IT outsourcing contracts suggests that modularity in KIBS presents peculiar features and difficulties. Indeed, IT outsourcing raises general questions about the coordination, control and boundaries of the large modern organization and about distributed innovation.

As suggested in the literature that matches product with organizational modularity, the rapid expansion of IT outsourcing in Germany and the UK can be partly explained by evidence of standardization of the 'rules of the games', in terms of standard contracts and the development of measurement and monitoring instruments. However, given the inseparability of production and information technologies, IT outsourcing is frequently accompanied by a more extensive transformation in the client firm's production technologies. The client therefore has an incentive to retain in-house capabilities to ensure coordination between IT and business strategy. IT outsourcing also demands firm-specific skills from transferred employees to facilitate organizational and knowledge coordination. Moreover, the intangibility of services exacerbates the tensions between client and supplier objectives, bringing into question the sustainability of innovation.

Our findings contribute to the literature on the limitations of the organizational implications derived from product modularity (Brusoni and Prencipe 2001; Prencipe 1997) and add a new limitation regarding problems for innovation in the context of design modularity in services. Because of the inseparability of information and production technologies and the intangibility of the services provided, IT outsourcing raises further questions about the innovation potential of modularity of KIBS. IT outsourcing has vastly increased the scale of computer services suppliers, suggesting that new organizational forms are central to economic growth founded upon certain KIBS. If the presumed positive effects of KIBS are to be realized, more attention needs to be paid to the interface between clients and suppliers – in particular, the role of the retained IT organization and transfer of staff – and to the mechanisms for reconciling conflicting interests and integrating knowledge between client and supplier. In their search for short-term cost-cutting, client organizations cannot delegate power and responsibility to computer service firms without jeopardizing their ability to upgrade innovation capabilities in the longer term. As such, understanding the dynamics of innovation in KIBS requires an understanding of the interplay between the development of capabilities within and across firms and knowledge and organization integration and coordination requirements.

ACKNOWLEDGEMENT

The authors are grateful for the financial support of the Anglo-German Foundation grant 1360, research assistance from Paulina Ramirez and the contributions of Matthias Knuth and Thorsten Kalina. We are especially grateful for the detailed comments of Mark Lehrer, which helped to reformulate the main argument.

NOTES

1. An early attempt to apply these ideas to services is Sundbo (2002). He suggests that in the long term, large firms in services may be moving into what he calls 'modulization' (which combines the productivity advantages of standardization and the individual customer satisfaction of customization) but finds only weak support for this in his empirical evidence.
2. KIBS include all business services that are knowledge-based, both based on social and institutional knowledge (as many of the traditional professional services) or more technological knowledge (e.g. computer, R&D and technical services). KIBS do not include many business services the roles of which are not knowledge-based (cleaning, catering, security, transport). In practice, however, there are many grey areas (Miles 2001).
3. As argued by Cohen and Levinthal (1990), a firm needs to develop 'absorptive capacity' in order that it may read its external technological environment and that it can draw upon that external knowledge as and when appropriate. This 'absorptive capacity' is largely a function of the firm's level of prior related knowledge.
4. As long as these rules are adhered to, designers can vary other aspects of the design. These aspects that can vary are 'hidden design parameters' because they do not need to be 'seen' by the designers of other parts of the system (Baldwin and Clark 2000).
5. Similarly, Jacobides and Winter (2003) combine the capability-based view of the firm and transaction costs theory to explore the effect of industry evolution on decisions of disintegration. In their argument, specialization and standardized information play an important role.
6. However, there are various accounts of the limitations of codification of knowledge (see the special issue of *Industrial and Corporate Change*, 9 (2) 2000).
7. 'Function points are a measure of the size of computer applications and the projects that build them. The size is measured from a functional, or user, point of view. It is independent of the computer language, development methodology, technology or capability of the project team used to develop the application' (SCT 1997).
8. In Cases 3 and 8, the multi-supplier contracts evolved accidentally due to mergers between the client and other organizations which had outsourcing contracts with alternative IT suppliers.

REFERENCES

Antonelli, C. (1998), 'Localized technological change, new information technology and the knowledge-based economy', *Journal of Evolutionary Economics*, **8** (2), 177–98.
Aoki, M. (1990), 'Towards an economic model of the Japanese firm', *Journal of Economic Literature*, **28** (1), 1–27.

Applegate, L. (1994), 'Managing in an information age: transforming the organization for the 1990s', in R. Baskerville, S. Smithson, O. Ngwenyama and J. DeGross (eds), *Transforming Organizations with Information Technology*, Amsterdam: North-Holland, pp. 15–96.

Arora, A. and A. Gambardella (1994), 'The changing technology of technical change: general and abstract knowledge and the division of innovative labour', *Research Policy*, **23** (5), 523–32.

Baldwin, C.Y. and K.B. Clark (1997), 'Managing in an age of modularity,' *Harvard Business Review*, **75** (5), 84–93.

Baldwin, C.Y. and K.B. Clark (2000), *Design Rules: The Power of Modularity*, Cambridge, MA: MIT Press.

Best, M. (1990), *The New Competition: Institutions of Industrial Restructuring*, Cambridge, MA: Harvard University Press.

Best, M. (2001), *The New Competitive Advantage: The Renewal of American Industry*, Oxford: Oxford University Press.

Brusco, S. (1982), 'The Emilian model: productive decentralisation and social integration', *Cambridge Journal of Economics*, **6** (2), 167–84.

Brusoni, S. and A. Prencipe (2001), 'Unpacking the black box of modularity', *Industrial and Corporate Change*, **10** (1), 179–205.

Brusoni, S., A. Prencipe and K. Pavitt (2001), 'Knowledge specialisation, organizational coupling, and the boundaries of the firm: why firms know more than they make?', *Administrative Science Quarterly*, **46** (4), 597–621.

Buck-Lew, M. (1992), 'To outsource or not to?', *International Journal of Information Management*, **12** (1), 3–20.

Chesbrough, H.W. and D.J. Teece (2002), 'Organizing for innovation: when is virtual virtuous?' *Harvard Business Review*, **80** (8), 127–36.

Cohen, W. and D. Levinthal (1990), 'Absorptive capacity: a new perspective on learning and innovation', *Administrative Science Quarterly*, **35** (1), 128–52.

Coombs, R. and S. Metcalfe (2000), 'Organizing for innovation: co-ordinating distributed innovation capabilities', in N. Foss and V. Mahnke (eds), *Competence, Governance, and Entrepreneurship: Advances in Economic Strategy Research*, Oxford: University Press, Oxford, pp. 209–31.

Cusumano, M.A. (1991), *Japan's Software Factories: A Challenge to US Management*, New York: Oxford University Press.

Daniels, P. and P. Moulaert (1991), *The Changing Geography of Advanced Producer Services: Theoretical and Empirical Perspectives*, London: Belhaven Press.

Davies, A. and T. Brady (2000), 'Organizational capabilities and learning in complex product systems: towards repeatable solutions', *Research Policy*, **29** (7–8), 931–53.

den Hertog, P. (2000), 'Knowledge intensive business services as co-producers of innovation', *International Journal of Innovation Management*, **4** (4), 491–528.

DiRomualdo, A. and V. Gurbaxani (1998), 'Strategic intent for IT outsourcing', *Sloan Management Review*, **39** (4), 67–80.

Earl, M. (1996), 'The risks of outsourcing IT', *Sloan Management Review*, **37** (3), 26–32.

Eisenhardt, K. (1989), 'Building theories from case study research', *Academy of Management Review*, **14** (4), 532–50.

Gallouj, F. and O. Weinstein (1997), 'Innovation in services', *Research Policy*, **26** (4–5), 537–56.

Gann, D.M. and A.J. Salter (2000), 'Innovation in project-based, service-enhanced firms: the construction of complex products and systems', *Research Policy*, **29** (7–8), 955–72.

Garud, R. and A. Kumaraswamy (1995), 'Technological and organizational designs for realising economies of substitution', *Strategic Management Journal*, **16** (special issue), 93–109.

Grandstand, O., P. Patel and K. Pavitt (1977), 'Multi-technology corporations: why they have "Distributed" Rather than "Distinctive Core" competences', *California Management Review*, **39** (4), 8–25.

Grant, R. (1992), 'The resource based theory of competitive advantage: implications for strategy formulation', *Sloan Management Review*, **33** (3), 114–35.

Henderson, R.M. and K.B. Clark (1990), 'Architectural innovation: the reconfiguration of existing product technologies and the failure of established firms', *Administrative Science Quarterly*, **35** (1), 9–30.

Henderson, J. H. and N. Venkatraman (1994), 'Strategic alignment: a model for organizational transformation via information technology', in T.J. Allen and M.S. Scott Morton (eds), *Information Technology and the Corporation of the 1990s*, Oxford: Oxford University Press, pp. 202–20.

Herrigel, G. (1993), 'Large firms, small firms, and the governance of flexible specialization: the case of Baden Wurttemberg and socialized risk', in B. Kogut (ed.), *Country Competitiveness: Technology and the Organization of Work*, Oxford: Oxford University Press, pp. 15–35.

Hobday, M. (2000), 'The project-based organization: an ideal form for managing complex products and systems', *Research Policy*, **29** (7–8), 895–911.

Holmstrom, B. and J. Roberts (1998), 'The boundaries of the firm revisited', *Journal of Economic Perspectives*, **12** (4), 73–94.

Huber, R. (1993), 'How Continental Bank outsourced its crown jewels', *Harvard Business Review*, **65** (1), 121–9.

Jacobides, M. and S. Winter (2003), 'Capabilities, transaction costs, and evolution: understanding the institutional structure of production', Reginald H. Jones Center working paper, The Wharton School, University of Pennsylvania.

Jonscher, C. (1994), 'An economic study of the information technology revolution', in T.J. Allen and M.S. Scott Morton (eds), *Information Technology and the Corporation of the 1990s*, Oxford: Oxford University Press, pp. 5–42.

Katsoulacos, Y. and N. Tsounis (2000), 'Knowledge-intensive business services and productivity growth: the greek evidence', in M. Boden and I. Miles (eds), *Services and the Knowledge-based Economy*, London: Continuun, pp. 192–208.

Krugman, P. (1999), 'The ascent of e-man', *Fortune* (May), accessed at http://web.mit.edu/krugman/www/whatsnew.html.

Lacity, M.C. and R. Hirschheim (1993), *Information Systems Outsourcing*, Chichester: Wiley.

Lacity, M.C. and R. Hirschheim (1995), *Beyond the Information Systems Outsourcing Bandwagon: The Insourcing Response*, Chichester: Wiley.

Lacity, M.C. and L. Willcocks (1994), 'Information systems outsourcing: a transaction cost interpretation of empirical evidence', OXIIM working paper, Templeton College, Oxford University, Oxford.

Lacity, M.C. and L. Willcocks (2001), *Global Information Technology Outsourcing*, Chichester: Wiley.

Lacity, M.C., R. Hirschheim and L. Willcocks (1994a), 'Realizing outsourcing expectations: from incredible expectations to credible outcomes', *Information Systems Management*, **11** (4), 7–18.

Lacity, M.C., L. Willcocks and D. Feeny (1994b), 'Information systems outsourcing: a decision making framework', OXIIM working paper, Templeton College, Oxford University, Oxford.

Langlois, R.N. (2002), 'Modularity in technology and organization', *Journal of Economic Behavior and Organization*, **49** (1), 19–37.

Langlois, R.N. (2003), 'The vanishing hand: the changing dynamics of industrial capitalism', *Industrial and Corporate Change*, **12** (2), 351–85.

Langlois, R. and P.L. Robertson (1992), 'Networks and innovation in a modular system: lessons from the microcomputer and stereo component industries', *Research Policy*, **21** (4), 297–313.

Langlois, R. and P.L. Robertson (1993), 'Business organization as a coordination problem: toward a dynamic theory of the boundaries of the firm', *Business and Economic History*, **22** (1), 31–41.

Langlois, R. and P.L. Robertson (1995), *Firms, Markets and Economic Change: A Dynamic Theory of Business Institutions*, London: Routledge.

Loh, L. and N. Venkatraman (1992), 'Determinants of information technology outsourcing: a cross-sectional analysis', *Journal of Management Information Systems*, **9** (1), 7–24.

Mahnke, V. (2001), 'The process of vertical dis-aggregation: an evolutionary perspective on outsourcing', *Journal of Management and Governance*, **5** (3–4), 353–79.

Mahnke, V., M.L. Overby and J. Vang (2003), 'Strategic IT outsourcing: what do we know and need to know?', paper presented at the DRUID Summer Conference 2003 'Creating, Sharing and Transferring Knowledge: The Role of Geography, Institutions and Organization', Copenhagen, 12–14 June.

Martin, J.A. and K.M. Eisenhardt (2003), 'Cross-business synergy: recombination, modularity and the multi-business team', Academy of Management 2003 Best Papers Proceedings.

Miles, I. (1993), 'Services in the new industrial economy', *Futures*, **25** (6), 653–72.

Miles, I. (2001), 'Knowledge-intensive business services revisited', Nijmegen Lectures on Innovation Management, Maklu, Antwerpen-Apeldoorn.

Miles, I. (2002), 'Services innovation: towards a tertiarisation of innovation studies', in J. Gadrey and F. Gallouj (eds), *Productivity, Innovation and Knowledge in Services*, Cheltenham, UK and Northampton, MA, USA: Edward Elgar, 164–96.

Miles, I. (2003), 'Business services and their contribution to their clients' performance: a review', contribution to ECORYS/CRIC project, CRIC, University of Manchester.

Miozzo, M. and I. Miles (2002), *Internationalization, Technology and Services*, Chelenham, UK and Northampton, MA, USA: Edward Elgar.

Miozzo, M. and L. Soete (2001), 'Internationalization of services: a technological perspective', *Technological Forecasting and Social Change*, **67** (2), 159–85.

Nelson, R.R. and S.G. Winter (1977), 'In search of useful theory of innovation', *Research Policy*, **6** (1), 36–76.

Ovum (2001), *The Holway Industry Report 2001: The Definitive Guide to UK Software and IT Services Companies and Markets*, London: Ovum.

Pavitt, K. (2003), 'Specialization and systems integration: where manufacture and services still meet', in A. Prencipe, A. Davies and M. Hobday (eds), *The Business of Systems Integration*, Oxford: Oxford University Press, pp. 78–94.

Peneder, M., S. Kaniovski and B. Dachs (2000), 'External services, structural change and industrial performance', Institute of Economic Research (WIFO) enterprise papers no. 3, Austria.

Piore, M. and C. Sabel (1984), *The Second Industrial Divide: Possibilities of Prosperity*, New York: Basic Books.

Prahalad, C. K. and G. Hamel (1990), 'The core competence of the corporation', *Harvard Business Review*, **68** (3), 79–91.

Prencipe, A. (1997), 'Technological competencies and product's evolutionary dynamics: a case study from the aero-engine industry', *Research Policy*, **25** (8), 1261–76.

Prencipe, A. (2000), 'Breadth and depth of technological capabilities in CoPS: the case of the aircraft engine control system', *Research Policy*, **29** (7–8), 895–911.

Quinn, J. (1992), 'The intelligent enterprise: a new paradigm', *Academy of Management Executive*, **6** (4), 44–63.

Sabel, C., G. Herrigel, R. Deeg and R. Kazis (1989), 'Regional prosperities compared: Massachusetts and Baden-Wurttemberg in the 1980s', *Economy and Society*, **18** (4), 374–404.

Sanchez, R. and J.Y. Mahoney (1996), 'Modularity, flexibility and knowledge management in product and organization design', *Strategic Management Journal*, **17** (Winter special issue), 63–76.

Sapir, A. (1987), 'International trade in services: comments', in O. Giarini (ed.), *The Emerging Service Economy*, Oxford: Pergamon Press.

Saxenian, A. (1994), *Regional Advantage: Culture and Competition in Silicon Valley and Route 128*, Boston, MA: Harvard University Press.

Scott Morton, M. S. (1991), *The Corporation of the 1990s: Information Technology and Organizational Transformation*, Oxford: Oxford University Press.

Silver, M. (1984), *Enterprise and the Scope of the Firm: The Role of Vertical Integration*, Oxford: Martin Roberston.

Software Computer Technologies (SCT) (1997), 'Function Point FAQ', accessed 4 October, 2003, at www.ourworld.compuserve.com/homepages/fpfaq.htm.

Steinmueller, W.E. (2003), 'The role of technical standards in co-ordinating the division of labour in complex system industries', in A. Prencipe, A. Davies and M. Hobday (eds), *The Business of Systems Integration*, Oxford: Oxford University Press, pp. 133–51.

Sturgeon, T.J. (2002), 'Modular production networks: a new american model of industrial organization', *Industrial and Corporate Change*, **11** (3), 451–96.

Sundbo, J. (2002), 'The service economy: standardisation or customisation?', *Service Industries Journal*, **22** (4), 93–116.

Teece, D. J. (1986), 'Firm boundaries, technological innovation and strategic management', in L.G. Thomas (ed.), *The Economics of Strategic Planning*, Lexington, MA: Lexington Books, pp. 187–99.

Tomlinson, M. (2001), 'A new role for business services in economic growth', in D. Archibugi and B.A. Lundvall (eds), *The Globalizing Learning Economy*, Oxford: Oxford University Press, pp. 97–110.

Utterback, J.M. (1994), *Mastering the Dynamics of Innovation: How Companies can Seize Opportunities in the Face of Technological Change*, Boston: Harvard Business School Press.

Venkatraman, N. (1991), 'IT-induced business reconfiguration', in M. Scott Morton (ed.), *The Corporation of the 1990s: Information Technology and Organizational Transformation*, Oxford: Oxford University Press, pp. 122–58.

von Hippel, E. (1994), 'Sticky information and the locus of problem solving: implications for innovation', *Management Science*, **40** (4), 429–39.

118   *Knowledge intensive business services and organizational forms*

Willcocks, L. and G. Fitzgerald (1994a), *A Business Guide to Outsourcing Information Technology: A Study of European Best Practice in the Selection, Management and Use of External IT Services*, London: Business Intelligence.

Willcocks, L. and G. Fitzgerald (1994b), 'Toward the residual IS organization?: research on IT outsourcing experiences in the United Kingdom', in R. Baskerville, S. Smithson, O. Ngwenyama, and J. DeGross (eds), *Transforming Organizations with Information Technology,* Amsterdam: North-Holland, pp. 129–52.

Williamson, O.E. (1975), *Markets and Hierarchies: Analysis and Antitrust Implications*, New York: Free Press.

Williamson, O.E. (1980), *The Economic Institutions of Capitalism*, New York: Free Press.

Wolff, E. (2002), 'How stagnant are services?', in J. Gadrey and F. Gallouj (eds), *Productivity, Innovation and Knowledge in Services*, Cheltenham, UK and Northampton, MA, USA: Edward Elgar, pp. 3–25.

Woolsey, J.P. (1994), '777', *Air Transport World*, **31** (4), 22–31.

Yin, R. (1981), 'The case study crisis: some answers', *Administrative Science Quarterly*, **26** (1), 58–65.

Yin, R. (1994), *Case Study Research: Design and Methods*, 2nd edn, Newbury Park, CA: Sage.

Young, A.A. (1928), 'Increasing returns and economic progress', *Economic Journal*, **38** (152), 527–42.

Zenger, T.R. and W.S. Hesterly (1997), 'The disaggregation of corporations: selective intervention, high-powered incentives, and molecular units', *Organization Science*, **8** (3), 209–22.

APPENDIX

Table A.4.1   Sources of interview data

Computer services firms		Interview data	Client organizations		Interview data
Name	Country of ownership	Job post of interviewee	Name	Sector	Job post of interviewee
Germany					
IBM	USA	Senior manager (Strategic Outsourcing)			
		Senior manager (Outsourcing Solutions)			
		Senior project manager (Hapag Lloyd)	Hapag Lloyd	Shipping	Senior manager (IT department)
		Senior project manager (Deutsche Bank)	Deutsche Bank	Banking	Senior manager (Technology Sourcing)
		Senior manager (Continental)	Continental	Manufacture of tires/rubber products	–
T-Systems	Germany	Senior executive manager			
		Senior project manager (Henkel)	Henkel	Chemicals	Senior manager (IT department)
		Senior project manager (Deutsche Telekom)	Deutsche Telekom	Communications	–
Logica	UK	Senior executive manager (German subsidiary)			
		Senior manager (Interbrew)	Interbrew	Brewer	Senior manager (IT department)

| Computer services firms | | Interview data | Client organizations | | Interview data |
Name	Country of ownership	Job post of interviewee	Name	Sector	Job post of interviewee
UK IBM	USA	Senior manager (Global Business Services)			
		Senior manager (Cable&Wireless)	Cable &Wireless	Communications	Senior manager (IT department)
		Senior manager (AstraZeneca)	AstraZeneca	Pharmaceuticals	–
EDS	USA	Senior manager (Client Relationships) Manager (Client Relationships)			
		Senior project manager (Inland Revenue)	Inland Revenue	Central gov't	Senior manager (IT department)
		Senior manager (DWP)	Department of Work and Pensions	Central gov't	–
		Senior manager (Rolls-Royce)	Rolls-Royce	Aerospace	Senior manager (Aerospace Operations)
		Senior executive manager (AT Kearney)			
Logica	UK	Senior manager (Sales) Senior manager (European market)			
		Senior manager (BP Chemicals)	BP Chemicals	Chemicals	Senior manager (Business Processes)
		Senior manager (Brittannia Airways)	Brittannia Airways	Airline	Senior manager (Information Systems)

5. Make and/or buy of IT-enabled services innovation: the case of the US express delivery industry

Volker Mahnke, Mikkel Lucas Overby and Serden Özcan

INTRODUCTION

IT-enabled innovations are of increasing importance for competitive success. By 'innovation' we mean 'the generation, acceptance, and implementation of new ideas, processes, products or services' (Thompson 1965, p. 2). Innovations are IT-enabled when they blend hardware and/or software assets with business capabilities to generate a novel process, product or service. If adopters of IT-enabled innovation do not command all necessary competence in-house and internal development is slow and costly, 'distributed capabilities' need to be coordinated across firm boundaries (Barney and Lee 2000; Coombs and Metcalfe 2000; Quinn 2000). Hence, firms must ally with external partners that can provide the required knowledge intensive business services through sourcing relations. Against the backdrop of the importance of IT-enabled innovation, the key question of this chapter is how do capability development strategies differ between first-movers and late entrants in IT-enabled services.

While the literature agrees on a general level that governance choices with regard to developing capabilities for IT-enabled services are consequential (Argyres and Liebeskind 2000; Barney and Lee 2000; Chesbrough and Teece 1996; Quinn 2000) because they have long-term consequences and are hard to reverse, it is far from clear how companies choose to develop capabilities for IT-enabled innovation – in-house or through outsourcing. On the one hand, Quinn (2000) asserts that today innovation calls for the complex knowledge that only a broad network of specialists can offer and that companies can profitably outsource almost any element in the innovation chain. Chesbrough and Teece (1996) by contrast argue when it comes to innovation, outsourcing inevitably produces more conflicts of interest than vertical integration, and those conflicts can hamper the kind of complex, systematic innovation that creates valuable business break-throughs. While the literature stresses the risks as well as the possibilities of sourcing IT capabilities, there seems to be a lack of an integrated and systematic analytical approach to the outsourcing decision when it comes to the development of capabilities conducive for IT-enabled innovation (Lacity and Hirschheim 1993).

This chapter develops theory based on three explorative in-depth case studies – FedEx, UPS and DHL – on why IT outsourcing strategies differ between innovative first-mover and late entrants seeking to adopt IT-enabled innovations. In particular, an analysis of three companies reveals that governance choices are influenced by a company's attempts to create, imitate, and/or leapfrog IT-enabled innovation in varying technological regimes. The remainder of this chapter proceeds as follows. First, we review the relevant literature to develop integrative propositions on the relationship between first- and late-moving firms and their outsourcing decision. We discuss our methodological approach to theory development and the context of our case research – the express delivery sector. Third, we comparatively discuss propositions and case findings to suggest implications for theory development and managerial practice. Conclusions follow.

CAPABILITY DEVELOPMENT FOR IT-ENABLED INNOVATIONS: MAKE AND/OR BUY

Several separate literature streams shed light on the question of why capability development strategies differ between companies adopting IT-enabled innovation as a first-mover and late-mover (Freeman and Soete 1997; Lieberman and Montgomery 1998). First-mover firms that pioneer the commercialization of IT-enabled innovation, such as online tracking systems, may be able to acquire a reputation as an industry leader; define the product/service category concept (for example, prototypicality) and shape buyer preferences for a product/service category; move down the learning curve fast to reduce cost; establish technical standards that late entrants are forced to follow; access superior consumer information under preferential uncertainty and differentiate offerings to segment the customers according to their willingness to pay (Carpenter, et al. 1997; Golder and Tellis 1993; Kerin et al. 1992; Porter 1983; Schmalensee 1982).

Another first-mover advantage exists if early adoption of IT-enabled innovation leads to buyer switching costs. Such switching costs are incurred for example when a novel IT-enabled service requires the structuring of client interfaces, clients have to learn the use of particular transaction platforms, and service clients make specific investments in IT systems. In addition, there are incentives for first-movers to build in incompatible systems design elements as this increases switching costs (Katz and Shapiro 1985) and to actively prevent the development of interface technologies that bridge otherwise incompatible technologies (Greenstein 1997). First-mover advantages, however, are only available if consumer adoption rates do not outpace attempts at competence development – either in-house or through outsourcing.

While first-movers may enjoy advantages through an early adoption of IT-enabled innovation, they also face considerable risks and often pay a substantial price for pioneering (Boulding and Christen 2001). A late adoption of IT-enabled innovation may provide late-mover advantages including the ability to free-ride on innovators' R&D through imitation, and making investments in technology infrastructure only after technological and market uncertainty have been resolved. In addition, late-movers are better positioned to perceive and exploit technological discontinuities that provide 'gateways' for leapfrogging. Late-movers' leapfrogging attempts are more likely to succeed, if various types of 'incumbent inertia' inhibit first-movers' adaptive response. For instance, technology sunk costs (Porter 1980), or inertia in the first-mover's processes, delays flexible adoption of capabilities (Lieberman and Montgomery 1988).

In the following section, we develop integrative propositions linking (a) the literature on firm boundaries and (b) the literature on the nature of technological change to outsourcing strategies of first- and late-movers.

Capability Development and the Boundaries of the Firm

Outsourcing during the adoption of IT-enabled innovation is broadly defined as a process undertaken by an organization to contract out the development of IT assets, staff and/or capabilities to a third-party supplier who in exchange receives monetary return over an agreed period of time (Kern et al. 2002). Primary theories explaining firm boundaries refer to production costs and transaction costs. Transaction cost economics views the firm as a contractual governance structure and stresses the transaction risks incurred in vendor relations, including unauthorized use of the firm's technology or know-how and ex-post extraction of rents generated through irreversible relationship-specific investments in capability development (Chesbrough and Teece 1996; Oxley 2000). In this view, outsourcing of capability development is constrained by contractual risks in vendor–client relations.

The knowledge-based view (Kogut and Zander 1992; Penrose 1959; Teece et al. 1997; Winter 1987) pictures the firm as a collection of knowledge assets and stresses production cost advantages that stem from specialization in particular routines and capabilities that cannot easily be imitated by competitors. Outsourcing of capability development conducive for IT-enabled innovation will be considered if vendors command comparative advantage in supplying capabilities cheaper and better (including the ability to act as transfer mechanisms for industry best practice) compared to the outsourcing firm. However, while outsourcing may help firms to access capabilities that they cannot build in a reasonable timeframe themselves, it also gives vendors a window to valuable knowledge that they may leak to other clients including competitors (Levin et al. 1987; Mansfield 1985).

Some authors argue that transaction cost economics and the knowledge-based view are complementary in the explanation of firm boundaries (Grant and Baden Fuller 2004; Langlois and Foss 1999; Madhok 2002). In sum then, companies will tend to rely on external partners in the development of IT capabilities, to the extent that (a) supplier competences are superior; (b) transaction risks are low; and (c) utilization of vendors does not pose severe imitation risks. Thus, we propose:

P1   Late-movers will outsource to a greater extent relative to first-movers because they face supplier markets that exhibit greater relative competence and higher competition between suppliers.

The Role of Technological Advancement

While current theories addressing firm boundaries give an indication of why outsourcing strategies differ between first-movers and late-movers, they are far from complete as they treat technological change and its implications – in terms of competition and economic organization – as exogenous. The literature on technological change is instrumental in developing more fine-grained propositions.

The performance of a technology has a recognized pattern over time, following an s-shaped diagram called the S-curve (Abernathy and Utterback 1978). In tandem with a technology's life cycle is its structural evolution. Most technologies evolve from an initial systemic phase to the opposite modular phase and then cycle back (Langlois and Robertson 1992). As the technology migrates from one phase to another, the optimal organizational configuration of the firm must also shift if it is to continue to capture value from its innovation (Chesbrough and Kusunoki 2001). When the technology is systemic, components are tightly integrated and inflexible. Component-system interfaces are poorly defined and ill-understood and are often in a state of flux. Consequently, IT engineers/designers encountering systemic architectures cannot, for instance, accurately measure important functionality attributes and do not yet understand how variation in one subsystem impacts on overall system performance (Christensen et al. 2002).

In order for performance improvements to be achieved, systemic architectures require intense 'unstructured technical dialogue' (Monteverde 1995) and 'iterative' (Von Hippel 1994) and 'overlapping' (Clark and Fujimoto 1991) problem-solving processes within the firm. Accordingly, the firm's IT development teams need to engage in direct observation; frequent face-to-face discussions, interaction with service prototypes through e.g. computer-based representation (Wheelwright and Clark 1992) and extensive learning by experimenting (Baldwin and Clark 1994). In addition, simultaneous development of various subsystems requires that efforts be closely paced to ensure the simultaneous attainment of development goals, which in turn necessitates close managerial monitoring and involvement (Teece 1994). At the systemic phase, vertical integration rather than the market constitutes the most efficient coordinating mechanism (Christensen et al. 2002; Monteverde 1995).

Gradually, however, the technology migrates towards modularity – giving way to standardization, codification and formalization of accompanying business processes (Worren et al. 2002). The decomposition of innovation's architecture encourages vertical specialization and leads to the establishment of networks of vendors with a standard of compatibility (Langlois and Robertson 1992). The presence of specialized component suppliers implies that firms need not source all components from a single vendor; instead they can mix and match components from multiple sources either in-house or through system integrators (Stremersch et al. 2003) even when the externally compatible components themselves carry systemic qualities (Langlois and Robertson 1992). Modularity then allows structured technical dialogue within and across the boundaries of the firm and thereby efficient vertical disintegration (Robertson and Langlois 1995). Thus we stipulate:

P2   Late-movers will outsource to a greater extent relative to first-movers because enabling IT is likely to be modular rather than systemic and structured technological dialogue allows for clear interface specification.

Two Types of Late-movers

While we expect that late-movers outsource more than first-movers, not all late-movers are alike. Accordingly, attempts at outsourcing capability development will differ[1] between two types of late-movers: (a) technology follower, and (b) technology leaders. Companies with weaker innovative capabilities are often forced to assume a late-mover position, and the best they can hope for is competitive parity with first-movers through successful imitation or external access of capabilities enabling IT services

(Cho et al. 1998; Hannan and McDowell 1987; Lieberman and Montgomery 1998). Accordingly, technology followers will tend to use external vendors extensively to access valuable capabilities if they contribute to fast and inexpensive imitation of a first-mover's service offerings. By contrast, firms with strong innovative capabilities may afford a 'wait-and-see' approach (Dasgupta 1988; Katz et al. 1985). As Freeman and Soete (1997, p. 273) explains some companies may:

> not wish to be the first in the world, but neither do they wish to be left behind by the tide of technical change. They may not wish to incur the heavy risks of being first to innovate and may imagine that they can profit from the mistakes of early innovators and from their opening up of the market.

In other words: the opportunity of learning from earlier moving competitors reduces incentives to immediate adoption (Chatterjee and Sugita 1990), and, given strong innovative capabilities, the chances of successful leapfrogging and imitation increase.

Distinguishing between types of late-movers is not only instrumental to identify imitation and leapfrogging threats that the first-mover is likely to encounter; they also expose the variation in late entrants' capability development approaches to the adoption of IT-enabled innovation. A late-moving technology follower will adopt a 'me-too' strategy with the objective of relying heavily on the technology work of first-movers, whereas technology leaders that move late do 'not normally aim to produce a carbon copy imitation of the IT-enabled innovation introduced by the early entrant' (Freeman and Soete 1997, p. 276). On the contrary, they will aspire to leapfrog to the extent their technological strength allows them to do so.

The literature on first-mover advantage posits that an imitative late-mover should enjoy decisive cost advantages to be able to match the lead held by the incumbent firms, provided that the technological advance remains undisrupted for an extended period (Lieberman and Montgomery 1998). Under such conditions, the longer the first-mover's lead-time is, the larger the late-mover disadvantages. Thus, speed to the market becomes highly critical for the imitating late-mover. For instance, Bowman and Gatignon (1996) and Ghosh et al. (1983) demonstrate that the order of entry (the late entrant being an imitator not an innovator) tends to diminish buyer response to price, quality and promotion, leading to the conclusion that imitating late entrants need to cut their prices and spend on promotion to a greater extent than the first-mover to achieve the same market share. This means the firm entering later has to emphasize speed of development and managerial, overhead and process efficiency (through, that is, standardization, flexibility and compatibility with existing technologies), as well as differentiated positioning (Carpenter and Nakamoto 1989; Christensen 1997) to offer low-cost (including transaction and learning costs) IT-enabled services. A competitive vendor market in this case will rapidly be able to provide the late-movers with low-cost IT-enabled system building and management due to scale, scope and learning economies derived from demand bundling and bulk purchase of off-the shelf IT tools and programs.

While imitation of best IT practices helps late-movers move up towards the industry's technology frontier, at best it leads to competitive parity not competitive advantage (Barney 1991) for which a late-moving technology leader is unlikely to settle. Empirical evidence shows that late-movers can overtake a pioneer through innovation since innovation in a product or service can reshape the corresponding prototypical category around which consumers form their preferences (Carpenter et al. 1997; Shankar et al. 1998). For a late entrant to leapfrog first-movers, at least the functionality of its IT-enabled service should exceed the value of that of first-mover's (Schilling 2003). Unless radical innovations are aimed at, leapfrogging rests primarily on the architectural innovations. For example, Christensen (1993) shows that firms entering the disk drive industry based on architectural innovation tended to perform much better than firms that entered based on component innovation. The distinction between component and architectural capabilities (Henderson and Clark 1990) is important for the outsourcing decision. Component capabilities concern the mastery of developing new functional components for an IT-enabled innovation system such as software and hardware. Architectural capabilities reconfigure such components with business capabilities in new ways. An architectural innovation then significantly changes the relation and interaction between components but leaves the components and inherent core design concepts unchanged. The implications for late-movers attempting imitation or leapfrogging are as follows:

P3 Late-movers seeking to imitate will outsource comprehensively – both architectural and component capabilities – compared to innovative late-movers, who will outsource selectively focusing on component capabilities.

THE EXPRESS DELIVERY MARKET AND IT-ENABLED INNOVATION

This section illustrates our propositions through an examination of make-and/or-buy decisions in the context of capability development for IT-enabled innovations. The research design is based on multiple cases in the same industry, allowing a replication logic whereby we used each case to test emerging theoretical insights (Yin 1989). This method allows for a close correspondence between theory and data, a process whereby the emergent theory is grounded in the data (Eisenhardt 1989; Glaser and Strauss 1967). This is an appropriate method for our purposes because we are engaged in an exploratory theory building exercise rather than theory testing. Cases serve as a context for theory-building, which can be extended to a wider context based on 'analytical generalization' (Yin 1989). We use secondary data from various sources including newspapers, annual reports, investment bank reports and general IT and business press outlets to provide an in-depth techno-history account of three leading express services firms operating in the USA in order of innovative entry: FedEx, UPS, DHL. The objective is to use the case studies as a specific context for theory development rather than to describe the firm comprehensively. Thus, we are selective, stressing empirical facts that are relevant to our theoretical argument on the interrelation between strategic postures and outsourcing of IT capabilities, thereby presenting only a partial picture of the complex companies. The setting is the US express delivery services market between 1984 and 2002. We choose the year 1984 as a point of departure as it signifies the date FedEx (then Federal Express) pioneered the electronic data interchange (EDI)-based shipping and tracking system. The solution quickly became the industry norm, maintaining its influence until the mid-1990s.

Industry Background

Express package delivery in the USA is a \$50 billion industry (2002 data) offering two basic products: overnight air and ground delivery. The former product is the more profitable of the two and largest in terms of market size. FedEx, United Parcel Services (UPS) and Airborne Express (acquired by DHL in 2003) dominate the domestic parcel market, which is the most important geographical region for all three companies generating over 80 per cent of their total revenues. In the air express market alone, FedEx, UPS and Airborne Express (DHL) control approximately 80 per cent market share. The only other large competitors in the US express market are the US Postal Service (USPS), Emery Express and the Brussels-based DHL Worldwide.

FedEx has been the market leader in the US express delivery services with a market share of 40 per cent or more since 1979. In fact the firm is generally credited for turning overnight delivery into a multi-billion dollar industry although the USPS pioneered the service with Express Mail in 1970. FedEx began transporting packages in April 1973. Volume picked up rapidly and service was extended. But the firm lost \$27 million in its first years; then in 1976 it earned $75 million in revenues. Its success quickly drew traditional cargo carriers into the new market. DHL entered the market in 1977, providing services on only a few selected domestic routes. Emery Air Freight, one of the largest domestic air-cargo carriers, followed in 1978 with the initial objective to carve out a 'heavyweight' niche (delivering heavy air cargo the next day) in the overnight services. Airborne Freight Corporation (DHL) entered in 1980 with a new name: Airborne Express. In 1982, the biggest of all private package delivery firms, UPS, moved into overnight delivery at prices that often were half of those charged by competitors. At the time of entry, UPS was the largest single private delivery firm on most railroads and owned the largest fleet of delivery trucks. Unlike Emery Air Freight and Airborne Express (DHL), which initially employed leased aircraft or made use of commercial airlines to ship parcels; UPS owned a large fleet of airplanes and operated out of its own hubs. As competition grew and the rapid build-up of air fleets created overcapacity, the average price of overnight delivery declined dramatically (for instance as of 1984, FedEx dropped its rate by 40 per cent over two years)[2] while the number of service offerings proliferated. The negative impact of fast-paced price cutting and discounting was soon augmented by the weakness in demand.[3] FedEx revenues rose by an average of 49 per cent during each of the years between 1976 and 1981. The firm reported a 33 per cent average increase in year-on-year revenues between 1982 and 1984. Table 5.1 details the overnight air market share between 1986 and 2001.

In-house Development of EDI-based Tracking Technologies: First-mover FedEx

In 1984, FedEx introduced the PowerShip Plus program. With the economic downturn and heightened competition, the timing of innovation was ideal. It reflected the founder and the CEO Frederick W. Smith's philosophy: To defend the firm's position against competitors by constantly innovating and improving in order to maintain a leadership position (Smith 1997). PowerShip Plus consisted of a DOS-based NEC PC with two printers located at the customer's site. The program provided over 25 000 high volume customers with proprietary online services including storing of frequently used addresses, labels printing, online package pick-up requests, package tracking, self-invoicing and report compilation. It was directly linked to the IBM mainframe on which FedEx's proprietary Customer Oriented Service and Management Operating System (COSMOS) ran. COSMOS was the firm's centralized management software, which managed delivery vehicles, packages and drivers while tracking weather shifts and even traffic jams. It was at the heart of FedEx IT architecture.

Table 5.1   US overnight air market shares

	Airborne (DHL) %	FedEx %	UPS %	USPS %	DHL %
2001	13	46	34	6	0.5
2000	13	47	34	6	0.4
1999	13	47	34	6	0.4
1998	13	47	32	6	0.3
1997	13	48	31	6	0.2
1996	13	49	32	6	0.2
1995	14	49	30	7	0.2
1994	13	50	30	7	0.2
1993	13	50	30	7	0.2
1992	12	50	30	8	1
1991	11	51	30	8	2
1990	10	51	31	8	2
1988	11	54	18	n.a.	3
1986	13	58	15	9	1

Note: Total figures do not add precisely due to rounding.

Sources: Healy (2003), Salomon Smith Barney (1999), Bear Stearns 15 October 1999, *Wall Street Journal* 5 July 1994, 8 January 1998.

Every customer service IT system interacted either directly or indirectly with COSMOS. Because PowerShip Plus was targeted at high-volume customers, FedEx paid for the system and its installation costs and provided customer training. The innovation fuelled FedEx's growth in the overnight express market. The program was so successful that, according to the CIO of FedEx, it quickly became one of the main distinctive value-added features of FedEx.[4] In 1986, analysts' estimates put FedEx with 58 per cent of the overnight market and followers UPS and Airborne (DHL) with 15 and 13 per cent respectively.

FedEx's technology pioneering was not a coincidence. Since its inception in 1973, the FedEx founder and CEO insisted that if FedEx were to compete with UPS – a firm that had existed for over half a century – a myriad of state-of-the art information systems had to be built alongside the air and vehicle networks. As early as 1979, Smith predicted the basis of competition in the next two decades – IT-enabled innovation. He argued that information about the package would soon be just as important as the delivery of the package and that the success of FedEx would be built on a bedrock of mobile computers, package-tracking systems, and sophisticated databases (Smith 1997). Hence, the firm soon began delivering breakthrough innovations. In 1979, FedEx launched the proprietary system COSMOS. COSMOS began as a dispatching system on a Burroughs machine in about 1976 but was rewritten (in Cobol) from scratch by FedEx software developers to run on IBM's special ACP operating system. In 1980, the firm released a proprietary wireless data network called Digitally Assisted Dispatch System (DADS) and became the first cargo carrier to adopt mobile data terminals and digital dispatch. DADS consisted of a central database (IBM 3081 mainframe), three call centres, 198 local dispatch stations, courier vans and the voice and data links between them. The system resulted in a 30 per cent increase in couriers' productivity the first day it was used. Then came the first version of PowerShip (an Epson system that operated like a postal meter) in 1981 and the PowerShip Plus in 1984 (Paul and Pearlson 1994). Smith saw it as a major breakthrough:

> To be able to deliver a new service you have to innovate. The hub-and-spoke distribution system which lies at the heart of the FedEx network is an example of that sort of innovation. Another way was the way we integrated ground and air systems from the very start. . . . Perhaps even more important was our recognition that, along with time-sensitivity, the ability to track the status of every item at every stage on its journey, from sender to recipient, would be crucial to customer satisfaction. We understood this even before we had the technological means to do it. As we have developed the means to do it, so information and IT have become central to the FedEx offer, next to our fleet of planes and trucks. (Smith 1997, pp. 217–18)

All IT-enabled innovations rested fundamentally on strong in-house application development capabilities and, as the *Infosystems* magazine observed, many were built 'in far less time than large corporations typically take to put together a garden variety of accounting systems'.[5] In 1983 alone, the firm spent around 5 per cent of its yearly revenue of $1 billion on IT, primarily on development initiatives. In 1985, FedEx employed 600 IT application developers (the entire IT department at rival UPS consisted of just 115 people) and was, as the only express courier firm, ranked among the top 100 corporate buyers of PCs by *InfoWorld*/Yankee Group. PowerShip was also a product of in-house experimentation. When FedEx sales people realized that the paperwork requirements of their largest customers were becoming very burdensome, they immediately met with the programmers from IT and came up with an interim solution; a lap-top supplied by FedEx running a simple DOS-based software program attached to a forms printer. The program used on-screen prompts to request shipping data. Frequent shipping destinations could be stored in a simple linked database. The FedEx management quickly realized the potential of the first PowerShip version and ordered the development of further built-in functions – above all, access to FedEx's IT-driven tracking system. The systemic nature of the project required a fluid organization with a high degree of instant information exchange on both the technical and the market sides. Accordingly, the development project began with the formation of cross-functional teams consisting of sales representatives and technical staff. All software developers were immediately assigned to major business functions to understand internal and external customer needs regarding shipping and tracking. To facilitate instant interaction and to improve problem solving, the management built a development lab, which from time to time hosted end users, as well. 'It (PowerShip) was originally designed to accommodate some very large shippers' said the then head of the development team, 'Once we got a handle on how to manage it, we realized that it probably had a lot of other applications. So we moved it out and just recently pretty much offered it to anybody that wants to use it'.[6]

Outsourced Development at the Innovative Late-mover UPS

UPS's entrance in the $2 billion express delivery market in 1982 posed significant challenges to the early entrants. As the CEO of FedEx stated: 'UPS will be in the overnight business for a long time. This is not a six-month drill'.[7] The Vice-Chairman of Airborne Express (DHL) also voiced concerns: 'No matter what area they go in they will have an impact because of their size and name recognition'.[8] Yet, in the subsequent two years, UPS had little direct effect on business. Now with PowerShip in place, FedEx executives were confident that their firm offered faster pick up, earlier delivery and superior ability to trace shipments than UPS.

Indeed until the late 1980s, UPS had traditionally relied on customers' confidence in their system to avoid providing tracking information to customers. As Langowitz (1992, p. 84) observed, 'To UPS, given their industrial engineering approach, tracking packages seemed to be an added expense with very little necessity – the system would deliver'. In an interview, the CEO, Kent Nelson, of UPS called many of the tracking systems overkill: 'It's not to imply that these are not excellent services, but there is a definite cost to them'.[9] FedEx CEO Frederick Smith offered an alternative perspective: 'The ability to track, trace and simply manage the large volumes of express items being moved will require automation and on-line integration of customer and carrier to an extent only barely discernible at present'.[10] The CEO of Airborne Express (DHL) acknowledged why it made sense to implement tracking and shipping systems:

> Just look at the growth pattern of shipping volume; it grew 45% last year [1985]. With that type of growth, a volume shipper can command big discounts.

The corporate shippers have become astute and are playing one courier against another. . . . One approach to combat the growing pressure is to offer 'non-pricing' enhancements, or offering high volume shippers computer time for printing air billing, in addition to package metering and monitoring shipping activity throughout the shipping process.[11]

In 1985, Airborne Express (DHL) followed FedEx in enabling large volume customers to access its online computer system FOCUS to check shipment status and obtain computer-to-computer invoicing services.[12] Despite relatively low-cost service offerings, UPS' entry had been far from desirable, as acknowledged by the CEO Kent Nelson: 'In 1983, we charged half the price for air delivery that Federal Express did and waited for customers to beat down our door. That didn't exactly happen'.[13] In late 1985 UPS acknowledged that the firm was lagging behind FedEx and Airborne (DHL) in IT. 'The thing we had to do to grow in [overnight] air [market] was to convince the shipping public [i.e. customers] that we could provide all the services that the leader – FedEx – has been able to provide. And that can only been done through technology' stated the UPS CEO Kent Nelson.[14] The firm had no automatic tracking system for its air shipments, a couple of mainframe computers and only 400 PCs. The CIO Franck Ebrick later recalled 'It was clear to us that we needed to change to meet customer needs to increase productivity in the electronic age. Federal Express had banks of old mainframes, which the young firm had grown with. But UPS, which took pride in personal service, was not yet plugged into the information age. . . . If you went into our information services facility in 1985, you went into 1975 in terms of technology'.[15] Led by the chairman, UPS in 1985 launched a five-year, $2 billion technology plan, with package tracking marking the entry point for IT infrastructure, as noted by the CIO Ebrick:

> There were some applications that were really critical to us, but tracking was always the Holy Grail. People were talking tracking this and tracking that. I said, 'Look, we don't have anything in place to do the tracking. We have no network. We have no database. We have no repository packaging. We are several years away from a sophisticated tracking system'. (Ross 2001, p. 3)

The build-up was immense. In 1983, UPS's Information Services (IS) Group, which was primarily dedicated to accounting, billing and operations reports, employed 90 people. In 1985, the same group comprised a mere 118 people and spent $40 million, 'a paltry figure for a corporation with $7.7 billion in revenue' wrote *Business Week*.[16] UPS started with 400 PCs in 1985, but by 1989, the firm had 20 000 PCs, five IBM mainframes, and opened up an $80 million computer and telecommunications centre, which linked all of UPS's computer networks. By 1991, UPS was able to boast a network – UPSNet – that linked six mainframes, approximately 250 mini computers, 40 000 PCs and 75 000 handheld units, connecting 1300 worldwide distribution sites.

While FedEx developed all its IT software in-house, UPS frequently turned to external suppliers to outsource component developments. Increasing vertical specialization within the technology and the rapid proliferation of customer demands meant that the company could enrol a more modular architecture. Initiating the restructuring exercise in 1985, UPS hired Andersen Consulting to help reconfigure data architecture plans. Subsequently, it teamed up with several external application development companies, including ConnectSoft (a specialist in Windows applications, electronic mail and on-line service), TanData (a logistics software solutions provider), Geographic Data Technology and MapInfo (digital mapping and tracking software developers). In designing the system that coded and tracked packages and automatically billed customers for customs duties and taxes, UPS also collaborated with Andersen's Management Information consultants. In addition UPS relied heavily on the consulting, software programming and training skills of Novell Networks. Just five years after FedEx's release of PowerShip Plus, UPS rolled out a matching system, called MaxiShip (in 1989). Similar to PowerShip Plus, MaxiShip came with a PC, printers and software that made it easier to create custom shipping manifests and management reports. Yet, it was based on a more modular architecture, which UPS executives viewed as superior to the competing offerings of FedEx: 'A lot of systems impose themselves, but with ours you can play around with the perimeter. . . . It has a different ethos behind its design. Those [FedEx] systems were created to make life easier for itself rather than its customers', noted the logistics systems manager at UPS, 'many companies have gone through the first generation of computers (that is, PowerShip and Easyship) and got upset with how inflexible they were. We have had to try and incorporate the sort of demands people have now and will have in the future'.[17] By 1993, about 8000 customers were handled through MaxiShip, equalling one third of FedEx's client base. The technology helped UPS enter the online services and narrow down the market/technology gap with FedEx. In 1988, the year before the launch of the UPS tracking system, FedEx controlled 54 per cent of the air-express market and UPS just 18 per cent; the following year, UPS's market share leapt to 31 per cent. In its quest to 'leapfrog not imitate the competitors' as the UPS's Kent Nelson put it,[18] UPS acquired Roadnet Technologies – a transport routing software developer and II Morrow – a maker of aviation and marine navigation systems whose vehicle-location technology was used by UPS in deploying more than 60 000 trucks in 1987. Both companies were highly instrumental in the development of tracking and shipping technologies.

Imitative Late-mover: Extensive Outsourcing by DHL

When DHL Worldwide appointed Michael Lanier as the new US CIO in 1990, his task was to rebuild the package handler's domestic information systems operations: 'DHL's US installation consists of a wide range of systems that don't talk to each other. . . . Even more serious is the firm's lack of a cohesive network architecture to link its domestic and international sites and support such strategic applications as package tracking, shipment control and customer service'.[19]

While FedEx was creating the express delivery industry in the USA during the 1970s, DHL was doing the same thing internationally.[20] By the mid-1980s, DHL emerged as the largest international delivery firm controlling over 50 per cent of the market. Nevertheless, DHL fell significantly behind in the growing US market by all measures, including market share and key applications such as package tracking. Moreover, FedEx and UPS invaded the international air-express market, leading to a 5 per cent drop in DHL's market share and eroding its profitability.[21] DHL tried to strengthen its presence in the USA through its prior working relationships with the majority of Fortune 500 companies overseas. In 1988, the firm moved its information systems activities into its US operating unit and in 1989, launched the EasyShip (DOS-based, written in Pascal) an integrated shipping processing system that allowed customers to have complete control in preparing and tracking shipments, all from their PC (similarly to Powership and Maxiship).

EasyShip quickly ran into operational challenges. At the root of the problem lay the UNIX system as it proved very difficult to incorporate the UNIX technology (mainframes bought from Pyramid Technology Corp) into the existing systems – mainframe and IBM's SSP and RPG II on 150 IBM System/36s.[22] Furthermore, the UNIX installation was neither robust nor cost-effective enough to support the highly strategic applications planned for launch, including customer service and package tracking.

After a brief evaluation of the current technology, the new CIO realized that the firm needed a state-of-the-art package tracking and handling system linked to a database and put together a revised architectural and technology plan. The plan entailed reengineering around a UNIX client/server system based on IBM RISC Syster/6000. DHL not only reintroduced an IBM mainframe to serve as a database repository, but also entered into a co-sourcing agreement whereby IBM acted as combined system integrator, co-developer and support provider. IBM received

$15 million over two years for completing the migration to the new system and developing and launching key applications such as package tracking and field systems. In 1994, DHL introduced an agent-based tracking service to communicate with US customers. DHL's system was based on software from Edify Corp and resided on IBM OS/2 servers.

Developing Web-based Tracking Technologies

Soon after came the Internet. In November 1994 FedEx pioneered the web-based tracking concept by launching fedex.com. The firm used Netscape clients, servers and development tools as a platform for building online applications and derived much of its online functionality from its Power-Ship package tracking software.[23] The website included a tracking feature that allowed customers to monitor their packages and is widely considered as one of the business world's first interactive web applications.[24] Jim Barksdale, former CIO and COO of FedEx, and then CEO of Netscape, says, 'It was the first outward and visible demonstration of a practical, productive use of the Internet by a real business for a real business purpose'.[25] Within ten months, approximately 17 000 people were tapping into the firm's web page daily of which 5500 were checking the status of delivery. The site exceeded the one million hits-per-month milestones at the end of 1995. By some estimates, FedEx was soon saving up to $4 million a year.[26] At the time, the founder and CEO Smith noted, 'Federal Express is just one enormous electronic neural system with 100 000 people and a few thousand trucks and planes and facilities appended to it – literally'.[27] The CIO Jones added, 'integration of Internet services with our transportation offerings is not an addition to our core business; it is our core business'.[28] FedEx reinforced its web presence with a number of Web-enabled innovations. In February 1995, the firm launched a limited beta test of FedEx Inter-NetShip, an application designed to allow customers to process a shipment from a web site. In late summer, a drop-box locator, enabling users to locate the closest of FedEx boxes, was added to the web site. In October 1996, it entered (first-mover among shipping firms) the Internet commerce services business with BusinessLink, a software and services package aimed at midsize businesses that sold products on the Web and would have them physically delivered by FedEx.[29] FedEx expanded the portfolio of services with the online catalogue and hosting system Virtual Order. Individual customers could also build integrated web sites using FedEx Applications Programming Interfaces. Using such systems allowed FedEx to encourage the growth of unique, content-driven web sites under merchant brands, which would also have FedEx capabilities integrated into them. In 1997, FedEx was receiving an average of 26 000 tracking requests a day and spent $1 billion on IT developments. In April 1998, the firm launched a redesigned fedex.com and the installation of the one-millionth customer electronic online shipping connection, which included FedEx PowerShip hardware/software shipping system. A year later fedex.com handled 60 million transactions per day. FedEx also used the Internet to refine its existing COSMOS system. Under the updated module, whenever a FedEx customer placed an order through fedex.com, the information went to COSMOS. The customer would then be able to track the status of the shipment through PowerShip. When initiating coverage on 22 June 1999, Citigroup's SmithBarney transportation analysts wrote:

> FedEx is determined to be the technology leader in the air express business and to embed itself in the Internet. Frankly we are impressed with what it has achieved in its technology offerings. From customs clearance to track and trace, Internet commerce strategies, and yield management, FedEx's technology is more advanced than anything we have seen elsewhere. We would note, however, that the technological gap with UPS has narrowed.

Indeed, the advent of the Internet also opened a new era in UPS' business. One month after FedEx established fedex.com, UPS created its own web page (December 1994). Although the initial web page was static, UPS quickly developed tools that would allow deeper integration into their customer's businesses and service systems. UPS quickly bundled its online services with an innovative set of application programming interfaces that let companies create their own hooks into functions such as package and signature tracking. Called, UPS OnLine Tools (consisting of UPS OnLine Professional, a windows-based system of package tracking and shipment processing; UPS OnLine Host Access, which links customers to a UPS data centre), they acted as the server side of an Internet client/server application (customers could set their e-commerce applications to act as clients to the UPS Online Tools while simultaneously acting as a server to end users' browsers). In 1996, the firm spent a huge portion of its $1.2 billion IT budget on development and expansion of these UPS online services. Its IT infrastructure included two data centres with nine mainframes, 250 minicomputers, 90 000 PCs, 77 000 portable computers, 2000 LANs, and 3000 dedicated lines.[30]

As in the previous technology cycle, the first-mover and innovative and imitative late-movers – FedEx, UPS and DHL, respectively – followed different approaches to IT development associated with internal versus external collaboration. In the words of *Business Week* magazine, '[while] FedEx forced customers to adopt its proprietary software . . . and shunned alliances until recently, UPS jumped into partnerships with giants such as Oracle and IBM'.[31] UPS had indeed become aggressive in establishing multiple relationships with e-commerce applications services providers like Open Market, Pandesic, SAP, Lotus, and NetDox Inc. and Tumbleweed software. UPS also established tie-ins with search engine vendors such as Yahoo. In contrast, FedEx argued that its 1,500 in-house programmers wrote more software code than almost any other non-software company.[32] 'We don't believe we are software developers', said UPS E-commerce chief, Mark Rhoney. Companies making that gamble 'are trying to go a bridge too far'.[33] UPS Vice Chairman and Executive VP Mike Eskew added:

> There were people out there, like IBM and Andersen and Harbinger, and hundreds of other folks that we've done alliances with, that did an awful lot of things better than we did. That really put us in the lead. Our competition wanted to do it all themselves, and that really gave us a leg up as we built these things. It was a good move for us. (Ross 2001, p. 4)

DHL joined the web fray almost eight months after the pioneer FedEx (in July 1995) with an IT budget outspent over four-to-one by FedEx and UPS.[34] The initial site was too static and did not offer any tracking service (in fact until 1997, DHL did not add any web-tracking service to the site and instead provided the same information through an integrated voice response system). Soon after the firm decided to change everything, the home page, linked pages, navigation and all links to legacy systems and initiated a reengineering project with the sole purpose of developing a web-centric infrastructure.[35] Even two years later, in 1997, the magazine *Information Week* wrote 'Three biggest shippers-FedEx, UPS and DHL let their customers track their packages over the Web. Of the three, DHL's Web-based service has been the weakest'.[36] When the first phase was completed in June 1998, DHL released its online shipment tracking system. Called DHL Connect, the system had a 'hybrid design', being partly Windows-based and partly Internet-based. It integrated client software with the World Wide Web. IBM helped create DHL Connect, including the development of the program in Java.

FedEx Moves to Outsourced Development

In 1998, the once unthinkable happened: UPS delivered about 55 per cent of all cyber shopping purchases, the US Postal Service (USPS) handled 32 per cent and FedEx captured only 10 per cent. FedEx was steadily losing market share to UPS even in segments where it had a commanding lead. For instance in the overnight service, UPS's volume grew faster than that of FedEx between 1996 and 2001. Besides, the Internet was driving new businesses to UPS as it was shipping for eBay and Amazon.com. Worse, not only intense competition but also IT itself in particular e-mail and fax was eating into FedEx's overnight delivery revenue. Most of the business documents that had traditionally been sent by overnight delivery were now being sent electronically. *Business Week* summed the situation up: 'UPS moved quicker into FedEx' turf than FedEx moved into that of UPS. And while Smith's early romance with computers gave him critical traction on the Internet, the technology is now undermining the choicest part of FedEx' operations: Overnight delivery, which makes up 50% of its revenues'.[37] FedEx was the first big transport firm to launch a web site with tracking and tracing capabilities but it failed to retain first-mover advantage as Smith acknowledged 'It was very clear to me that this [Internet] was going to change the whole way that people interacted with each other. What I didn't understand was how rapidly it would be adopted'.[38]

These developments increasingly pushed FedEx to outsource as obtaining competences required producing, enhancing and maintaining IT-enabled innovations. In an unusual bid, the firm let its internal IT work with SAP to develop a shipping application, formed an alliance with Interworld Corporation – a provider of enterprise-class e-commerce software systems-and hired Lokion Interactive as content manager of the web site, fedex.com. In 2001, FedEx signed a development deal with wireless application provider W-Technologies to offer customers the ability to wirelessly track shipments, determine the status of a shipment and e-mail that information to multiple e-mail accounts.

DISCUSSION

The following discussion assesses our propositions against the case findings (see Table 5.2). While the evidence supports our propositions, the analysis of outsourcing during the adoption of IT-enabled innovation at FedEx, UPS, and DHL suggests three additional themes for theory development. The first relates to systematic differences in the decision situation between first- and later-movers along three dimensions: adoption risks, supplier competence, and transaction risk. If adoption risks are high – that is, consumer's speed of switching from one particular service offering to an IT-enabled service offering is unknown – first-movers considering outsourcing need to trade-off R&D risk-sharing with vendors and the simultaneously increased likelihood of imitation via vendors. Late-movers do not face such trade-off as their objective is to match or leapfrog first-mover advantages. Here external supply of component capabilities does not increase competitive risks and by implication increases the likelihood of sourcing IT competences externally.

Table 5.2 Propositions and case-study findings

	Propositions	FedEx	UPS	DHL
P1	Late-movers will outsource to a greater extent relative to first-movers because they face supplier markets that exhibit greater relative competence and higher competition between suppliers.	FedEx was the first entrant into the overnight package services market. It was also the first firm to offer IT-enabled tracking and shipping services to the customers. By handing out free hardware and software and providing customer education, FedEx converted the market into an electronic one. Yet, it did so through strong in-house IT capabilities.	UPS lacked the infrastructure and competences to internally build and implement the tracking system and hence immediately turned to the market. Yet it wanted to leapfrog rather than imitate FedEx, and thus had to develop in-house competences in the process of outsourcing. To that end, the firm acquired two software development firms. By the end of 1980s, the US vendor market was well developed and EDI had gained acceptance.	DHL also chose to rely on vendor market expertise, as the firm did not have the necessary IT muscle in the USA. At the same time, it ran into significant infrastructure problems that it was unable to solve internally. DHL did not face any imitation risks.
P2	Late-movers will outsource to a greater extent relative to first-movers if enabling IT is modular rather than systemic because structured technological dialogue allows for clear interface specification.	The project was rather unstructured initially but once its potential was realized, FedEx management formed cross-functional internal teams as it needed to define new functionalities and improve	The enabling IT was more modular than four years previously when FedEx built its tracking and shipping application. UPS chose multiple contracting approaches with a number of technology vendors, which	DHL too benefited from modularity and in fact one of the first decisions was to overhaul the closed and inflexible infrastructure in favour of a more modular structure.

performance. Unstructured technical dialogue was key for the innovation enhancement.

P3 Late-movers seeking to imitate will outsource comprehensively – both architectural and component capabilities – compared to innovative late-movers, which will outsource selectively focusing on component capabilities.

provided various components of the system. It followed open standards.

Leapfrogging the pioneer was the objective of management which meant that innovation along functionality was the key as FedEx had built a sizeable advantage based on its three-year head start and its reputation for creating the express package delivery. UPS chose selective outsourcing of components while retaining the architectural development of the system in-house.

DHL sought to imitate rather than expand the technology frontier in the industry. One reason was that it was serving on a few routes for some 2500 large customers. DHL hired IBM as a system integrator and system developer. In a two-year contract, IBM renewed the infrastructure and developed major applications.

First-movers, who consider outsourcing, need to account for unspecified interfaces in the early phase of technology development due to unstructured technological dialogue. Late-movers do not face such difficulties if technology has moved from the integral to the modular phase. Simultaneously, due to unstructured technological dialogue, transaction risks are higher for first-movers compared to late-movers. By contrast late-movers can control for transaction risks because structured interfaces allow for more complete contracting with multiple competent suppliers (Poppo and Zenger 1998). By implication, late-movers face lower transaction risks and can take full advantage of vendors' comparative advantage unconstrained by complications of unstructured technological dialogue and other transaction risks (Christensen et al. 2002). In addition, due to limited vendor scale, scope, and learning economies in the early phase of market development, first-movers considering outsourcing are constrained in finding competent and specialized vendors. Late-movers face such difficulties to a lower degree. In sum, differences in decision-making parameters between first- and second-movers substantiate our assertion that boundary choices differ accordingly.

The second theme explores the interplay between decision makers' uncertainty over the nature of technological advance. Technological volatility increases decision makers' uncertainty over whether technological advance is competence-destroying or enhancing. First-movers considering outsourcing need to trade-off performance disadvantages of being locked into obsolete technology (where technological change is destructive) with differentiation advantages (where technological change is enhancing). As shown in the case of FedEx, perceptions of technological change and its impact are coloured by the path-dependent history of internal capability development: Misperceptions of technological change are the result of myopia leading to competence traps (Levinthal and March 1993).

Late-movers seek to minimize the risk of being locked in obsolete technology and to recognize early on any new technological developments that the first-mover may overlook. For example, imitative late-movers will make new partnering arrangements only after technological uncertainty and demand uncertainty has resolved, as the DHL case illustrates. In addition, as the successful leapfrogging attempt by UPS shows, being engaged in a wide net of diverse external knowledge sources helps avoid competence traps. Late-movers attempting leapfrogging, provided they command innovative capabilities, are better positioned to recognize technological breakthrough, and can tailor timing of vendor partnering accordingly.

The final theme explores the inter-dependence of boundary decisions and the creation and defence of first-mover advantages. Our evidence shows that varying degrees of outsourcing have implications for creating and defending first-mover advantages. FedEx, the first-movers in both EDI and web-enabled tracking systems – in the beginning of a technology life cycle – refrained from extensive outsourcing for three reasons. First, supplier markets remained underdeveloped (Willcocks and Fitzgerald 1994) and technological uncertainty increased the risk of contractual failures (Williamson 1975). In addition, extensive outsourcing would have increased the risk of imitation by latecomers leading to competitive parity. While the first-mover initially confronts interdependent interfaces, where 'unstructured technical dialogue' occurs, moving along a particular technology life cycle, first-movers will tend to in-source (outsource) early on (later on) IT-related component processes as the technology life cycle proceeds and marginal improvement possibilities level out. Interestingly, first-movers facing technological discontinuity will encounter performance penalties if integrated in obsolete competences (Afuah 2001), as will first-movers outsourcing too late in the development of a particular life cycle. By implication, the timing of outsourcing arrangements and their antecedence as well as the avoidance of first-mover's cognitive biases constitute a crucial area of future research to actively pursue.

While the three themes discussed above inform theory development that moves beyond the current discussion in the literature, they also have managerial implications. The practical implications of this chapter are to move beyond simplified recommendations in the literature either stressing risks (Chesbrough and Teece 1996) and possibilities of outsourcing (Quinn 2000) in the context of adopting innovations respectively (cf. the discussion of transaction cost economics and the knowledge-based view respectively). While classifying risk and benefits of outsourcing remains important, not all benefits and risks are equally relevant for first- and late-movers in the adoption of IT-enabled innovation. Recognizing the role of technological advance (competence enhancing vs. competence destroying) and attempted strategic posture (first-mover, imitative or leapfrogging late-mover) mediates both risks and available benefits in important ways.

CONCLUSION

This chapter has argued that IT-enabled innovations are of increasing importance for competitive success in a range of industries including express delivery services. How companies choose to develop associated competences – in-house and/or through outsourcing – is consequential for creating and sustaining competitive advantage. The key concern of this chapter was to address the crucial question: how do capability development strategies differ between first-movers and late entrants in IT-enabled services? We developed theory based on three explorative case studies – FedEx, UPS and DHL, and the analysis of these companies revealed that governance choices are influenced by a company's attempts to innovate, imitate, and/or leapfrog IT-enabled innovation in varying technological regimes. Importantly, the nature of technological advance, as well as attempted strategic posture, influence transaction costs and possibilities of tapping into comparative advantages of outsourcing vendors in the adoption of IT enabled innovations.

NOTES

1. Central to this variation is the firm's 'strategic orientation', which bears a significant impact on its entry timing decision (Schoenecker and Cooper 1998). As strategic orientation is not merely a function of capabilities and resource profiles but also organizational attributes, firm history and management attitudes, later entry does not necessarily always imply comparatively weak innovative capabilities and a lack of critical resources on the part of entrant.
2. *Washington Post*, 4 November 1984, 'Overcoming time at Memphis hub'.
3. *New York Times*, 2 March 1983, 'The air delivery free-for-all'.
4. *Infoworld*, 16 November 1992, 'Federal Express gives clients on-line tracking system'.
5. *Infosystems*, May 1985, 'Redefining an industry through integrated automation'.
6. *Transportation and Distribution*, June 1988, 'EDI just arriving in air express industry'.
7. *New York Times*, 2 March 1983, 'The air delivery free-for-all'.
8. *New York Times*, 2 March 1983, 'The air delivery free-for-all'.
9. *New York Times*, 9 June 1985, 'United Parcel extends its reach'.
10. *Air Transport World*, November 1985, 'Federal Express: big, bigger, biggest'.
11. *Washington Business Journal*, 24 February 1986, 'Delivery services, couriers compete for growing market', **4** (41), 16–17.
12. *Administrative Management*, October 1985, 'Overnight and faster'.
13. *Chief Executive*, March 1994, 'The wizard is OZ: interview with UPS CEO Kent Nelson'.
14. *Information Week*, 18 May 1992, 'Tracking technology'.
15. *Across the Board*, 17 February 1989, 'Technology helps delivery giant expand and fend off competitors'.
16. *Business Week*, 25 August 1988, 'UPS gets a big package – of computers'.
17. *Computer Weekly*, 12 December 1991, 'United picks a package to pass the parcels around'.
18. *Business Week*, 9 November 1987, 'Why Federal Express has overnight anxiety'.
19. *Computerworld*, 17 September 1990, 'Time to take control at DHL'.
20. *Business Week*, 9 May 1983, 'An international courier takes on Federal Express'.
21. *Forbes*, 21 September 1987, 'Downdraft'.
22. *Computerworld*, 10 April 1989, 'DHL sold on Cosmopolitan Unix'.
23. *PC Week*, 1 August 1996, 'FedEx adds shipping to web'.
24. *Information Week*, 20 November 1995, 'Powering up: FedEx CIO takes overnight delivery to the next level'.
25. *Internetweek*, 25 October 1999, 'FedEx delivers on CEO's IT vision'.
26. *PC Week*, 1 November 1996, 'FedEx adds shipping to web'.
27. *Internetweek*, 25 November 1999, 'FedEx delivers on CEO's IT vision'.
28. *Internetweek*, 25 November 1999, 'FedEx delivers on CEO's IT vision'.
29. *Information Week*, 14 October 1996, 'FedEx delivers web commerce'.
30. Information Week, 25 November 1996, 'UPS expands online package'.
31. *Business Week*, 21 May 2001, 'UPS vs. FedEx: ground wars'.

32. *Wall Street Journal*, 20 November 1998, 'Will FedEx shift from moving boxes to bytes'.
33. *Wall Street Journal*, 4 October 1999, 'Ante up: big gambles in the new economy'.
34. *Computerworld*, 15 June 1998, 'Look out, Goliath, David has a rock'.
35. *Computerworld*, 24 February 1997, 'Keeping up with the Joneses'.
36. *Information Week*, 2 June 1997, 'DHL's biggest package'.
37. *Business Week*, 21 May 2001, 'UPS vs. FedEx: ground wars'.
38. *Business Week*, 21 May 2001, 'UPS vs. FedEx: ground wars'.

REFERENCES

Abernathy, W.J. and J.M. Utterback (1978), 'Patterns of industrial innovation', *Technology Review*, **80** (7), 40–47.

Afuah, A. (2001), 'Dynamic boundaries of the firm: are firms better off being vertically integrated in the face of a technological change?', *Academy of Management Journal*, **44** (6), 1211–28.

Argyres, N. and J. Liebeskind (eds) (2000), *The Role of Prior Commitment in Governance Choice*, Oxford: Oxford University Press.

Baldwin, C.Y. and K.B. Clark (1994), 'Modularity-in-design: an analysis based on the theory of real options', Harvard Business School working paper, Cambridge, MA.

Barney, J. (1991), 'Firm resources and sustained competitive advantage', *Journal of Management*, **17** (1), 99–120.

Barney, J. and W. Lee (eds) (2000), *Multiple Considerations in Making Governance Choices: Implications of Transaction Cost Economics, Real Option Theory, and Knowledge Based Theories of the Firm*, Oxford: Oxford University Press.

Bear Stearns & Co. Equity Research (1999), Airborne Freight Corporation, 15 October.

Bear Stearns & Co. Equity Research (2001), ' FedEx Corp', 8 May.

Boulding, W. and M. Christen (2001), 'First mover disadvantage', *Harvard Business Review*, **79** (9), 20–1.

Bowman, D. and H. Gatignon (1996), 'Order of entry as a moderator of the effect of marketing mix on market share', *Marketing Science*, **15** (3), 222–42.

Carpenter, G.S. and K. Nakamoto (1989), 'Consumer preference formation and pioneering advantage', *Journal of Marketing Research*, **26** (3), 285–98.

Carpenter, G.S., D.R. Lehman, K. Nakamoto and S. Walchli (1997), 'Pioneering disadvantage: consumer response to differentiated entry and defensive imitation' working paper, Northwestern University.

Chatterjee, R. and Y. Sugita (1990), 'New product introduction under demand: uncertainty in competitive industries', *Managerial and Decision Economics*, **11** (2), 1–12.

Chesbrough, H. and K. Kusunoki (eds) (2001), *The Modularity Trap: Innovation, Technology Phase Shifts and the Resulting Limits of Virtual Organization*, London: Sage Publications.

Chesbrough, H. and D. Teece (1996), 'When is virtual virtuous? Organizing for innovation', *Harvard Business Review*, **74** (1), 65–73.

Cho, D.-S., D.-J. Kim and D.K. Rhee (1998), 'Latecomer strategies: evidence from the semiconductor industry in Japan and Korea', *Organization Science*, **9** (4), 489–505.

Christensen, C.M. (1993), 'The rigid disk drive industry: a history of commercial and technological turbulence', *Business History Review*, **67** (4), 531–88.

Christensen, C. (1997), *The Innovator's Dilemma – When New Technologies Cause Great Firms to Fail*, Boston, MA: Harvard Business School Press.

Christensen, C.M., M. Verlinden and G. Westerman (2002), 'Disruption, disintegration and the dissipation of differentiability', *Industrial and Corporate Change*, **11** (5), 955–93.

Clark, K.B. and T. Fujimoto (1991), *Product Development Performance: Strategy, Organization and Management in the World Auto Industry*, Boston, MA: Harvard University Press.

Coombs, R. and S. Metcalfe (eds) (2000), *Organizing for Innovation: Coordinating Distributed Innovation Capabilities*, Oxford: Oxford University Press.

Dasgupta, P. (1988), 'Patents, priority and imitation or, the economics of races and waiting games', *Economic Journal*, **98** (389), 66–80.

Eisenhardt, K. (1989), 'Building theories from case study research', *Academy of Management Review*, **14** (4), 532–50.

Freeman, C. and L. Soete (1997), *The Economics of Industrial Innovation*, London: Pinter.

Ghosh, A., S.A. Neslin and P.W. Schoemaker (1983), 'Are there associations between price elasticity and brand characteristics', proceedings of the American Marketing Association, August.

Glaser, B.G. and A.L. Strauss (1967), *The Discovery of Grounded Theory – Strategies for Qualitative Research*, New York: Aldine de Gruyter.

Golder, P.N. and G.J. Tellis (1993), 'Pioneer advantage: marketing logic or marketing legend?', *Journal of Marketing Research*, **30** (2), 158–70.

Grant, R.M. and C. Baden Fuller (2004), 'A knowledge accessing theory of strategic alliances', *Journal of Management Studies*, **41** (1), 61–84.

Greenstein, S.M. (1997), 'Lock-in and the costs of switching mainframe computer vendors: what do buyers see?', *Industrial and Corporate Change*, **6** (2), 247–73.

Hannan, T.H. and J.M. McDowell (1987), 'Rival precedence and the dynamics of technology adoption: an empirical analysis', *Economica*, **54** (214), 155–71.

Healy, P. (2003), 'United Parcel Services IPO', HBS case no. 9-103-015.

Henderson, R. and K. Clark (1990), 'Architectural innovation: the reconfiguration of existing product technologies and the failure of established firms', *Administrative Science Quarterly*, **35** (1), 9–30.

Katz, M.L. and C. Shapiro (1985), 'Network externalities, competition, and compatibility', *American Economic Review*, **75** (3), 424–40.

Kerin, R.A., P.R. Varadarajan and R.A. Peterson (1992), 'First-mover advantage: a synthesis, conceptual framework, and research propositions', *Journal of Marketing*, **56** (October), 33–52.

Kern, T., L. Willcocks and E. Heck (2002), 'The winner's curse in IT outsourcing: strategies for avoiding relational trauma', *California Management Review*, **44** (2), 47–69.

Kogut, B. and U. Zander (1992), 'Knowledge of the firm, combinative capabilities, and the replication of technology', *Organization Science*, **3** (3), 383–97.

Lacity, M.C. and R. Hirscheim (1993), 'The information systems outsourcing bandwagon', *Sloan Management Review*, **35** (1), 73.

Langlois, R.N. and N.J. Foss (1999), 'Capabilities and governance: the rebirth of production in the theory of the firm', *Kyklos*, **52** (2), 201–18.

Langlois, R. and P.L. Robertson (1992), 'Networks and innovation in a modular system: lessons from the microcomputer and stereo component industries', *Research Policy*, **21** (4), 297–313.

Levin, R.C., A.K. Klevorick, R.R. Nelson and S.G. Winter (1987), 'Appropriating the returns from industrial research and development; comments and discussion', *Brookings Papers on Economic Activity*, **3**, 783–832.

Levinthal, D.A. and J.G. March (1993), 'The myopia of learning', *Strategic Management Journal*, **14** (special issue, Winter), 95–112.

Lieberman, M.B. and D.B. Montgomery (1988), 'First-mover advantages', *Strategic Management Journal*, **9** (special issue, Summer), 41–58.

Lieberman, M.B. and D.B. Montgomery (1998), 'First-mover (dis)advantages: retrospective and link with the resource-based view', *Strategic Management Journal*, **19** (12), 1111–25.

Madhok, A. (2002), 'Reassessing the fundamentals and beyond: Ronald Coase, transaction cost and resource-based theories of the firm and the institutional structure of production', *Strategic Management Journal*, **23** (6), 535–50.

Mansfield, E. (1985), 'How rapidly does new industrial technology leak out?', *Journal of Industrial Economics*, **34** (2), 217–23.

Monteverde, K. (1995), 'Technical dialog as an incentive for vertical integration in the semiconductor industry', *Management Science*, **41** (10), 1624–38.

Oxley, J. (ed.) (2000), *Learning Versus Protection in Inter-firm Alliances*, New York: Palgrave Macmillan.

Paul, D.L. and K.E. Pearlson (1994), 'Federal Express: the role of IT in customer service', case study, University of Texas at Austin.

Penrose, E. (1959), *The Theory of the Growth of the Firm*, Oxford: Oxford University Press

Poppo, L. and T. Zenger, (1998), 'Testing alternative theories of the firm: transaction cost, knowledge-based, and measurement explanations for make-or-buy decisions in information services', *Strategic Management Journal*, **19** (9), 853–77.

Porter, M.E. (1980), *Competitive Strategy*, New York: Free Press.

Porter, M.E. (ed.) (1983), *The Technological Dimension of Competitive Strategy*, Greenwich, CT: JAI Press, Inc.

Quinn, J.B. (2000), 'Outsourcing innovation: "The new engine of growth"', *Sloan Management Review*, **41** (4), 13–28.

Robertson, P.L. and R. Langlois (1995), 'Innovation, networks and vertical integration', *Research Policy*, **24** (4), 543–62.

Ross, J. (2001), 'United Parcel Services: delivering packages and E-commerce solutions', MIT working paper.

Salomon Smith Barney Equity Research (1999), 'FDX Corp./Airborne Freight initiating coverage', 22 June.

Schilling, M.A. (2003), 'Technological leapfrogging: lessons from the US video game console industry', *California Management Review*, **45** (3), 6–32.

Schmalensee, R. (1982), 'Product differentiation advantages of pioneering brands', *American Economic Review*, **72** (3), 349–65.

Schoenecker, T.S. and A.C. Coops (1998), 'The role of firm resources and organizational attributes in determining entry timing: a cross-industry study', *Strategic Management Journal*, **19**, 1127–43.

Shankar, V., G. Carpenter and L. Krishnamurthi (1998), 'Late mover advantage: how innovative late entrants outsell pioneers', *Journal of Marketing Research*, **35** (1), 54–70.

Smith, F. (ed.) (1997), 'Federal Express: the supremely packaged warehouse in the Sky', in F. Gilmose (ed.) *Brand Warriors*, London: HarperCollinsBusiness, pp. 217–28

Stremersch, S., A.M. Weiss, B. Dellaert and R.T. Frambach (2003), 'Buying modular systems in technology-intensive markets', *Journal of Marketing Research*, **40** (3), 335–50.

Teece, D. (ed.) (1994), *Design Issues for Innovative Firms: Bureaucracy, Incentives and Industrial Structure*, Oxford: Oxford University Press.

Teece, D., G. Pisano and A. Shuen (1997), 'Dynamic capabilities and strategic management', *Strategic Management Journal*, **18** (7), 509–33.

Thompson, V.A. (1965), 'Bureaucracy and innovation', *Administrative Science Quarterly*, **10** (1), 1–20.

Von Hippel, E. (ed.) (1994), ' "Sticky information" and the locus of problem solving: implications for innovation', *Management Science*, **40** (4), 429–39.

Wheelwright, S. and K. Clark (1992), *Revolutionizing Product Development*, New York: Free Press.

Willcocks, L. and G. Fitzgerald (1994), *A Business Guide to IT Outsourcing*, London: Business Intelligence.

Williamson, O.E. (1975), *Markets and Hierarchies*, New York: Free Press.

Winter, S.G. (ed.) (1987), *Knowledge and Competence as Strategic Assets*, Cambridge, MA: Ballinger.

Worren, N., K. Moore and P. Cardona (2002), 'Modularity, strategic flexibility and firm performance: a study of the home appliance industry', *Strategic Management Journal*, **23** (12), 1123–40.

Yin, R.K. (1989), *Case Study Research-Design and Methods*, London: Sage.

PART II

Knowledge intensive business services in diverse national contexts

6. Institutional effects on the market for IT outsourcing: analysing clients, suppliers and staff transfer in Germany and the UK

Damian Grimshaw and Marcela Miozzo

INTRODUCTION

This chapter explores the influence of the institutional context on the development of a fast-growing area of knowledge intensive business services (KIBS) – the market for IT outsourcing – in Germany and the UK. KIBS are defined as those industries that have a relatively high skilled workforce and/or are intensive users of high technology (OECD 1999; p. 18). Several studies suggest that KIBS play a valuable role in economies as intermediary inputs (Daniels and Moulaert 1991; EC 1997), although their contribution is strongly influenced by the nature of client–supplier relations (Miles 2003; Tomlinson 2001). Nevertheless, few comparative studies have examined the role of institutional processes in shaping KIBS sectors and there has been very little attention to how these processes impact upon both supplier and client organizations.

Investigation of institutional processes extends the results of existing literature in three potential ways. First, comparative analysis of IT outsourcing in Germany and the UK illuminates the interplay between heterogeneous institutional processes that impact upon IT outsourcing and the pressures for convergence exercised by powerful computer services firms. The 'varieties of capitalism' literature demonstrates that economic behaviour is embedded in institutions, guided by shared values and subject to path-dependent change (Hall and Soskice 2001), and that differences in organizational performance result in part from country-specific institutional effects (Lehrer and Darbishire 1997; Vitols 2001). Nevertheless, the potential for 'common systemic trends' is stronger in sectors dominated by powerful multinational corporations (MNCs) and characterized by rapid technological change (Strange 1997). Our study seeks to contribute to our knowledge of these twin processes by extending varieties of capitalism approach to computer services and the IT out-sourcing market.

Second, a focus on the client and the supplier illuminates how management capabilities on both sides of the IT outsourcing contract are influenced by institutions and market structure, and, in turn, how these capabilities shape the quality of services provision. Client capabilities are an important factor in achieving performance gains in interactions with provider firms (Earl 1996; Gallouj 1997), but to date there has been little cross-fertilization with results from other studies that identify the institutional influences on inter-organizational relations (Arrighetti et al. 1997; Lane and Bachmann 1997). Our analysis focuses on two areas where institutional processes impact upon client–supplier outsourcing arrangements: management of staff transfer; and client expertise.

Third, because knowledge assets – in the form of human and intellectual capital – are central to the expansion of KIBS sectors (Teece 1998), our study seeks to illuminate how institutionally embedded employment relations mutually interact with new organizational forms. While study of changing organizational forms has, in recent decades, been largely neglectful of work and employment issues (Barley and Kunda 2001), new empirical investigations demonstrate that the management of a range of human resources issues are central to the dynamics of new organizational forms (Marchington et al. 2005; Swart and Kinnie 2003). Our research addresses these issues by exploring the role of staff transfer from client organizations to IT firms.

The market for IT outsourcing was selected for research because in recent years it has underpinned growth in the computer services sector and is characterized by highly complex, long-term contractual agreements between clients and suppliers (Lacity and Willcocks 2001). Like other KIBS sectors, it involves intensive use of high technologies and specialized skills and knowledge, and is considered vital as a source of job growth and value-added (Miles 2003). Many KIBS depend to some extent on practices of externalizing business activities from other organizations (Coe 2000); IT outsourcing provides an extreme example as market growth results from organizations replacing in-house provision with services provided by specialized computer services firms. Expansion thus appears contingent upon institutions that can foster and support highly adaptable organizational structures. Exploration of evidence for Germany and the UK provides a good test of this notion, since one might expect the need for fast-changing organizational structures to conflict with the slow, incremental process of change characteristic of Germany's 'deliberative' (Hall and Soskice 2001) institutions; by contrast, strong managerial discretion, high labour mobility and evidence of radical organizational restructuring in the UK (Ackroyd and Procter 1998) suggest a good 'fit' with the apparent deregulatory bias of the IT outsourcing market.

The nature and heterogeneity of institutional effects are investigated in this study through comparative analysis of 13 large IT outsourcing contracts involving staff transfer – six in Germany and seven in the UK – using quantitative industry-level data and qualitative interview data from client organizations and computer services firms. The 13 contracts were selected from a sample of four computer services firms – two US, one German and one UK-owned. Our first aim was to analyse industry-level trends in each country in order to address the proposition that Germany suffers from a 'service gap' in the evolution of the IT outsourcing market. Drawing on qualitative data, the second aim was to assess the influence of 'deliberative institutions' on the nature and pace of development of IT outsourcing in each country. The third aim was to identify those factors that facilitated client organizations in managing complex IT outsourcing contracts, including institutional effects on technical knowledge and contract expertise, as well as organizational practices related to the specificities of IT outsourcing and the highly concentrated base of suppliers.

The chapter begins by drawing a bridge between the findings from studies on KIBS and cross-national evidence of the importance of country institutions in shaping inter-firm relations. Section 3 identifies three areas of institutional effects on country markets for IT outsourcing, namely, the compatibility of the institutional context for market growth, the impact of 'deliberative institutions' on the pace and trajectory of IT outsourcing and the effect of systems of legal rules and technical standards on client capabilities. Section 4 describes the research methodology. Section 5 presents the empirical evidence and section 6 discusses the results in the context of findings and theoretical issues in related studies.

KNOWLEDGE INTENSIVE BUSINESS SERVICES AND INTER-ORGANIZATIONAL RELATIONS

There are strong claims in a growing number of studies that KIBS play an important role in the economy, especially as innovative users of new technologies and motors for employment growth (EC 2003). The evidence since the late 1970s suggests a strong expansion in their share of the total economy in OECD countries as a result of radical organizational and technological change (Peneder et al. 2003) and points to their important role as intermediate inputs, improving competitiveness in other sectors (Miles 2001). However, while cross-national comparisons demonstrate differences in the role of KIBS in different national economies (Antonelli

1998; Tomlinson 2001), there has been little detailed exploration of the effects of the institutional context in structuring markets for KIBS. Institutions are potentially important given that the evolution of these services sectors depends very much on complex, often long-term, interactions with client organizations in other sectors of the economy (Miles 2003). Several studies have sought to understand how these relations impact on the evolution of different areas of KIBS but have to date tended to rely on the use of broad taxonomies (e.g. Tordoir's (1995) use of jobbing, sparring and sales relations), rather than detailing the influence of institutional and organizational characteristics of different sectors and different economies.

The notion that there is likely to be an important role for institutional effects in structuring client–supplier relations finds support in a separate literature on country differences in the character of inter-firm relations. Differences in the degree of cooperation in inter-firm relations are found to be shaped by societal institutions, such as trade associations, employer associations, trade unions and government legislation (Lane and Bachmann 1997), as well as by informal social norms (Arrighetti et al. 1997) and the technological and economic context (Sako 1992). Comparative evidence for Germany and the UK, for example, shows that inter-firm relations 'are constituted within the specific contexts of social institutions which channel the expectations and actions of buyers and suppliers' (Lane and Bachmann 1997, p. 250). A focus on inter-firm relations in a single market – that is, IT outsourcing – extends this literature by illuminating the interactions between institutional effects, the peculiarities of large IT outsourcing arrangements (complex long-term contracts, staff transfer and the retention of in-house client capabilities) and the dominance of multinational computer services suppliers.

A focus on a single sector responds to the need to explore how varieties of capitalism may be elaborated across a range of dimensions, with the potential for convergence differing by sector (Hollingsworth et al. 1994). Because IT plays a critical role in strategy formulation and implementation in all organizations (Venkatraman 1991), and is non-separable from many production technologies and new modes of services delivery (Jonscher 1994), institutions that develop and support coordination between client and IT supplier play a critical role. However, the idiosyncratic nature of the supply base, with a strong representation of multinational computer services firms (Coe 2000), means that a country's institutions may not 'fit' with policies and practices adopted by the leading firms in the sector (Strange 1997), with consequent strong pressures towards convergence in inter-organizational arrangements across countries shaped by the diffusion of what Djelic and Quack (2003) call the 'challenger rules' of multinational companies.

INSTITUTIONAL EFFECTS ON THE MARKET FOR IT OUTSOURCING IN GERMANY AND THE UK

Analyses of the institutional frameworks of capitalism illuminate important differences between the German and UK systems of business, employment and innovation (Crouch and Streeck 1997; Lane 1989; Marsden 1999; Nelson 1993; Whitley 1999). In Germany, organizations operate within a 'coordinated market economy' (Hall and Soskice 2001), constituted by a regulatory approach to business law, strong networks of trade associations, high participation by organizations in industry associations, low penetration of financial markets, a highly reputable system of vocational training and a consensual approach to industrial relations – underpinned by a 'dual system' of trade unions and legally mandated works councils (Culpepper and Finegold 1999; Hyman 2001; Tylecote and Conesa 1999). The UK, by contrast, is an example of a 'liberal market economy'. Organizations operate within a highly 'financialized' economy, a deregulated labour market characterized by weak trade unions and fragmented employer and trade associations, a tradition of adversarial industrial relations and an underdeveloped system of vocational training (Cully et al. 1999; Froud et al. 2000).

Appreciation of these differences suggests three areas where the interlocking institutions of Germany and the UK are likely to shape the development and form of IT outsourcing. At the most general level of analysis, there is a debate concerning the 'goodness of fit' between a country's institutions and the development of business services, with some studies arguing for the greater 'comparative institutional advantage' (Hall and Soskice 2001) of Anglo-Saxon economies (Streeck and Heinze 1999, cited in Bosch 2001). While Germany occupies an excellent position in the international technology race – it has one of the highest contributions of R&D intensive industries to GDP (Meyer-Krahmer 2001) – there are concerns about its ability to adapt to new fast-changing markets. Challenges include the need to transform the public research infrastructure to fit modern modes of knowledge production, increase the flexibility of actors to 'shift' the research frontier and develop the services sector to boost job growth (ZEW 1999). Also, Germany's slow and coordinated institutional processes, with a strong emphasis on codified procedural rights, have been found to impede radical restructuring within organizations (such as redefining core activities or entering new markets) (Lehrer and Darbishire 1997; Vitols 2001). Overall, its particular institutional features are said to make it especially difficult for Germany to enter fundamentally new technological trajectories like IT (ZEW 1999), which may impair the development of KIBS and hinder economic growth.

In the UK the debate is different. The service economy is highly developed and, unlike Germany, organizations have been quick to reverse strategies of internalization in favour of externalization, encouraging rapid growth in business services sectors dependent on outsourcing (Colling 2000). However, there are concerns regarding the sustainability of network-based business services provision, due to doubts about how benefits are distributed (Coombs et al. 2003) and difficulties in managing employment (Marchington et al. 2005). Indeed, while KIBS are relatively developed in the UK, their contribution to GDP is low compared to other countries (Tomlinson 2001). One reason is the nature of UK institutional conditions underpinning client–supplier relations, such that 'the promotion of the unfettered expansion of these types of services without any consideration for the economic network of which they form a part would not necessarily be beneficial' (Tomlinson 2001, p. 103). Therefore, with regard to the IT outsourcing market, what is needed is a comparative assessment of its development in Germany and the UK in order to 'test' the compatibility of each country's institutional context with this particular business services market.

A second more specific institutional effect on the market for IT outsourcing concerns the strength of 'deliberative institutions', which provide actors with an opportunity for collective discussion and for reaching agreement (Hall and Soskice 2001, p. 12). Decisions to enter into or exit from supplier contracts involve the interests of the different organizations party to the contract, as well as workers who may face transfer of employers where in-house activities are outsourced. Deliberation thus provides actors with an opportunity to establish, through common agreement, the risks and gains associated with the new venture and to address distributive issues. While such procedures may underpin long-term stability and cooperation, they may also delay responsiveness, resulting in a laggard trajectory of organizational change or innovation development.

Deliberation is stronger in Germany than in the UK. Managers in German organizations do not have a capacity for unilateral action, as they do in the UK, since they must secure agreement with supervisory boards, which involve employee representatives, and, typically, representatives from banks and major suppliers (Jackson et al. 2004). Also, the mandatory system of works councils in Germany provides workers with rights to company information, consultation and co-decision-making; veto rights cover personnel policies on working time, careers, dismissal, payment systems and in-house training (Slomp 1995). By contrast, companies in the UK are not required to establish works councils, granting managers considerably greater prerogative in the design and implementation of organizational policy. Where organizational strategies of externalization involve staff transfer to a supplier organization, German organizations must undertake extensive deliberation, based on a legal requirement for negotiation and consultation with worker representatives. Where outsourcing involves 'company transfer' (*Betriebsübergang*) and/or 'company change' (*Betriebsänderung*), then civil law code (article 613a, BGB, *Bürgerliches Gesetzbuch*) and legal regulations on workplace labour relations (*Betriebsverfassungsgesetz*) establish a role for works councils to protect the rights of transferring workers (EIRO 2001). Works councils have rights to consultation about outsourcing, and workers are protected against dismissal, have the right to refuse to transfer on an individual basis, and enjoy protected terms and conditions. Only the latter condition applies in the UK, as derived from European legislation on the Transfer of Undertakings Protection of Employment, which protects workers' terms and conditions of employment at the point of transfer. This condition is weaker than in Germany since article 613a extends protection for 12 months after transfer. Hence, German workers are better protected than in the UK and employers spend more time with works councils designing a consensual outsourcing process. In exploring the influences of deliberative institutions on the IT outsourcing market, there is a need to examine their effects on the pace and trajectory of change in inter-organizational arrangements.

A third institutional effect on the IT outsourcing market concerns the way different country systems of legal rules and technical standards shape the capabilities of client and supplier organizations to design and manage an effective interchange of services. We observed above that the value-added of KIBS depends as much on the expertise of the client as that of the supplier organization (Gallouj 1997; Miles 2003). Studies of consultancy relations suggests that clients need to have the capabilities to select among competing suppliers, to combine externalized activities with in-house functions, to negotiate respecifications of long-term contracts and to understand (and possibly control) future strategic developments (McGivern 1983; Sturdy 1998). While there is some focus on the role of social networks and organizational cultures in shaping client capabilities (Hislop 2002), we need to consider other studies on inter-organizational relations for insights into institutional effects.

A country's system of contract law (as used to adjudicate incomplete contracts) has been found to influence the ability of contracting organizations to establish cooperative relations (Lane and Bachmann 1996). Germany's 'regulatory approach' to contract law reflects broader societal norms of fairness through utilizing the norm of 'good faith' (*Treu und Glauben*, as enshrined in article 242 of the German *Bürgerliches Gesetzbuch* – Zimmermann and Whittaker 2000). In particular, German courts have wide powers to police standardized contracts and make adjustments where there is inappropriate delegation of risk by the more powerful party (Casper 2001, p. 390). The overall approach is sustained through a positive feedback effect through the 'collective law-making capability' of German firms, which make high use of standardized contracts based on legal and technical rules set out in industry frameworks (Casper 2001, p. 393). Empirical evidence suggests that these collectively developed industry frameworks generate widespread acceptance of legal rules among purchasing and sales managers (Arrighetti et al. 1997), which in turn has a trust-enhancing effect between business partners (Lane and Bachmann 1996). Contract law in the UK primarily seeks to protect each party's freedom to contract and to enforce the written contract. There is no general rule requiring parties to conform to good faith (although similar results are sometimes achieved through a piecemeal approach – Zimmermann and Whittaker 2000) and UK law contains fewer mechanisms to protect the weaker party in circumstances where unanticipated events impose an imbalance in bargaining power between parties (Goode 1995). Its evolution lacks a conscious design due to the historical suspicion about statutory intervention, thus restricting the scope for legal regulation of inter-firm relations (Atiyah 1995). Compared to Germany, empirical evidence suggests more limited legal expertise in British firms (Stewart et al. 1994, cited in Lane 1997) and less acceptance and knowledge of the law among managers responsible for contractual arrangements with other firms (Arrighetti et al. 1997).

Technical standards, as institutionally constituted within a particular country, also shape the contracting capabilities of client and supplier organizations. Technical standards improve market transparency and facilitate market transactions between buyers and sellers (Tate 2001). Also, since standard setting is typically a collective effort, it encourages greater acceptance and knowledge of technical norms, which reduces the costs of writing and interpreting job tasks specified in inter-organizational contracts and reduces the risk of conflict among contracting partners (Hölmstrom 1985; Lane 1997). The strong role of technical standards associated with German inter-organizational relations (Tate 2001) fits with a cultural emphasis on technical skill, reflected by the reputable credentialist systems of vocational training and the high status accorded to technically qualified people (Steedman and Wagner 1989). By contrast, low technical knowledge among UK managers – reflecting the weak adoption and diffusion of standards, in part due to low membership of trade associations (Lane 1997) – raises the question as to whether UK client organizations lack the expertise to manage contracts with IT supplier firms, thus failing to benefit fully from advances in technology. In Germany, however, the consensual development of industry-wide standards may also carry the risk of what Lane (1997) calls 'cognitive lock-in', where technological change proceeds along established paths, reducing scope for radical innovations.

In seeking to explore the extent and heterogeneity of these different institutional effects on the IT outsourcing market in Germany and the UK, our study focused on the characteristics of the market in each country and on the policies and practices deployed by client and supplier partner organizations engaged in IT outsourcing contracts. As such, we sought to identify heterogeneous institutional effects through qualitative data that illuminated effects within the organizational setting.

METHODOLOGY

This research collected qualitative data through case studies of four computer services firms and a 'case survey' (Yin 1981) of 13 IT outsourcing contracts. The case survey approach is recommended where certain isolated factors are worthy of substantive attention (Yin 1981, p. 62). Here, our aim was to complement the four case studies with a detailed focus on the nature and form of the IT outsourcing contract. Thus, a case survey of contracts was designed to include interviews with senior managers from the client and IT firm responsible for each contract. The research also drew on secondary quantitative data for the IT outsourcing market in Germany and the UK and organization- and contract-related documentation. The research was undertaken between September 2002 and September 2003.

The key benefits of the chosen method are its sensitivity to complex heterogeneous circumstances, its capacity to facilitate exploratory discovery (enabling interplay between data collection and analysis) and its suitability for analysing inter- and intra-country patterns among the four case studies and the different outsourcing contracts, with presentation of data in tabular form (Eisenhardt 1989; Yin 1994). There are also limitations. The choice of a case survey approach rather than detailed case studies of a smaller number of IT outsourcing contracts generates a risk that there is insufficient information on the unique, contextually-specific nature of each outsourcing contract (Miles 1979). Also, the number of factors worthy of examination may be large relative to the number of contracts selected and there is a risk that single factors are compared across contracts in a way that oversimplifies the context (Yin 1981). Nevertheless, our chosen research strategy provides a clear focus on a single market (IT outsourcing) and fills a gap in the literature on KIBS by exploring cross-national differences in both client and supplier practices.

The selection of IT outsourcing contracts in Germany and the UK proceeded by first negotiating access to four leading computer services firms that (a) had signed large IT outsourcing contracts and (b) reflected a mix of country of ownership. Two firms (US-owned IBM and UK-owned Logica) agreed access in both Germany and the UK. US-owned EDS was visited in the UK and German-owned T-Systems in Germany. Data from initial background interviews were supplemented by a wealth of secondary source data from company reports, trade journals (e.g. *Computer Weekly*), industry consultants (Ovum in the UK and Pierre Audoin Consultants in Germany) and government reports (e.g. UK Public Accounts Committee).

Analysis of initial data generated a sample of client organizations, which held large IT outsourcing contracts with the selected computer services firms. For each firm in each country we sought access to senior project managers on the client and supplier sides of between two and three IT outsourcing contracts, in order to control for unique cases within each firm. On the supplier side, we achieved this in all firms except Logica in Germany (where we only explored one IT outsourcing contract) and on the client side we interviewed managers in eight out of a possible 13 contracts (Table 6.1b). In three contracts the project manager in the IT firm advised us that access was not possible in the partner client organization because the commercial relationship was too sensitive. In the two contracts involving IBM in the UK, access to the client was agreed in principle but due to restructuring within the client firms actual contact proved impossible. We selected clients from those sectors of economic activity that generated demand for IT outsourcing. As we show below (next section), these were different in each country, with a strong role of the public sector in the UK and greater demand from manufacturing in Germany. While some of the contracts involved operations across more than one country, cross-border contracting was not a feature of any of the 13 contracts investigated. The sample was thus not random, but reflected the selection of specific contracts in key sectors to extend theory on institutional effects to a broad range of inter-organizational contexts (Eisenhardt 1989, p. 537).

Two types of interviewees were chosen. In the computer services firms we sought interviews with senior and executive managers with responsibilities for IT services to gain detailed information on firm strategy (Table 6.1a). Also, in both the client organization and the computer services firm, we interviewed the senior project manager responsible for managing the IT outsourcing contract in order to identify potentially diverging perspectives. Themes addressed included the rationale for IT outsourcing, contract details, staff transfer, direct and indirect performance gains to the client, inter-organizational relations and examples of tensions and conflicts. All 31 interviews were semi-structured and lasted one to two hours. They were followed up by telephone conversations and e-mails to request further

Table 6.1 Sources of interview data

(a) General interviews at the four computer services firms

Computer services firm	Country of ownership	Job post of interviewee	Country of operation
IBM	USA	Senior manager (Strategic Outsourcing)	Germany
		Senior manager (Outsourcing Solutions)	Germany
		Senior manager (Global Business Services)	UK
EDS	USA	Senior manager (Client Relationships)	UK
		Manager (Client Relationships)	UK
		Senior executive manager (AT Kearney)	UK
T-Systems	Germany	Senior executive manager	Germany
Logica	UK	Senior executive manager	Germany
		Senior manager (Sales)	UK
		Senior manager (European market)	UK

(b) Interviews for 13 IT outsourcing contracts

Computer services firm	Job post of interviewee	Client organization	Job post of interviewee	Country of operation
IBM	Senior project manager (Hapag Lloyd account)	Hapag Lloyd (shipping)	Senior manager (IT department)	Germany
	Senior project manager (Deutsche Bank account)	Deutsche Bank (banking)	Senior manager (Technology Sourcing)	Germany
	Senior manager (Continental account)	Continental (tyre manufacture)	No interview	Germany

Table 6.1 (continued)

(b) Interviews for 13 IT outsourcing contracts

Computer services firm	Job post of interviewee	Client organization	Job post of interviewee	Country of operation
	Senior manager (Cable&Wireless account)	Cable&Wireless (communications)	No interview	UK
	Senior manager (AstraZeneca account)	AstraZeneca (pharmaceuticals)	No interview	UK
EDS	Senior project manager (Inland Revenue account)	Inland Revenue (central govt.)	Senior manager (IT department)	UK
	Senior manager (DWP account)	Department of Work and Pensions (central govt.)	No interview	UK
	Senior manager (Rolls-Royce account)	Rolls-Royce (aerospace)	Senior manager (Aerospace Operations)	UK
T-Systems	Senior project manager (Henkel account)	Henkel (chemicals)	Senior manager (IT department)	Germany
	Senior project manager (Deutsche Telekom account)	Deutsche Telekom (communications)	No interview	Germany
Logica	Senior manager (Interbrew account)	Interbrew (brewer)	Senior manager (IT department)	Germany
	Senior manager (BP Chemicals account)	BP Chemicals (chemicals)	Senior manager (Business Processes)	UK
	Senior manager (Brittannia Airways account)	Brittannia Airways (airline)	Senior manager (Information Systems)	UK

information. At the end of the project, all participants received a summary report and had the opportunity to provide feedback. In agreement with the participants, we name the organizations investigated but preserve confidentiality in the presentation of qualitative data.

Over the 12-month period of data collection, there was a continuous iterative process between the analysis of interview notes and documentary information and the revision of themes for subsequent data collection (Miles and Huberman 1994). IT outsourcing contracts investigated at the early stages of the project tended to require more follow-up contact to provide answers to questions that emerged later in the research process. All interviews were taped and transcribed. Rather than analyse the qualitative data with computer software packages (Weitzman and Miles 1995), we adopted the old-fashioned approach of identifying themes from a direct reading of transcripts. All data were coded and tabulated manually (available on request). The benefit was both to provide a check between the data and our knowledge of the organizational and institutional context and to reveal findings that did not necessarily fit with initial expectations.

THE EMPIRICAL EVIDENCE

This section presents the research evidence around the three issues identified in the above analysis of institutional effects:

1. The need for a comparative assessment of the nature and development of the IT outsourcing market in Germany and the UK;
2. The influence of deliberative institutions on the pace and trajectory of change in inter-organizational arrangements; and
3. Identification of the institutions (legal rules/technical standards) and organizational practices that strengthen client expertise in negotiating and managing IT contracts.

Throughout the discussion, IT outsourcing is defined as multi-year contracts that involve the transfer of assets (infrastructure and staff) from a client organization to the supplier, as well as the shift of managerial control and responsibility for service delivery.

Comparing the IT Outsourcing Markets

The market for IT outsourcing developed later in Germany than in the UK. For most of the 1990s the UK led the European IT outsourcing market. Following a 'shakeout' in the computer services industry – associated with a shift by mainframe manufacturers to provide computer services (IBM, ICL, Compaq), and the growth of software firms (Logica, Oracle, Microsoft) and outsourcing firms (EDS, Sema) (Coe 2000) – coupled with a trend among UK organizations to divest non-core activities, the UK witnessed average annual growth rates of 40 per cent in the IT outsourcing market between 1991 and 1996 (Ovum 2002, p. 218). By 1996 the outsourcing and processing market constituted 20 per cent of the total UK software and computer services industry (Ovum 2001, Figures 2.6.4, 2.12.1). In Germany, IT outsourcing only significantly began to develop after the mid-1990s. Prior to this, most IT outsourcing was 'captive outsourcing'; for example, more than 80 per cent of IT outsourcing revenue of Debis Systemhaus came from its parent company, Daimler Chrysler (PAC 2003, Figure 2.1). After 1993–94, Germany's 'non-captive' market expanded significantly. This coincided with the entry of IBM, with its strategic goal of increasing outsourcing revenue in Germany, and the take-off of Germany's largest software firm, SAP, which developed a system for standardizing software applications, making the German landscape suddenly more conducive to IT outsourcing (Lehrer Chapter 7 this volume).

During the early 2000s, Germany and the UK contributed most to the growth in the European IT outsourcing market (Table 6.2). By 2002, the UK and German IT outsourcing markets accounted for well over half of the total Western European market, €31 billion out of a total €53 billion. In both countries, growth in IT outsourcing outstripped growth in the total industry for software and IT services (Figure 6.1).

Despite similarity in size and growth rates, there are important country differences. First, there remains a significant captive market in Germany (defined to include only those suppliers that have contracts with their parent company and contracts with external clients), estimated at €4200 million in 2002 (33 per cent of the total IT outsourcing market), compared to €0.5 million in the UK in 2000 (PAC 2003; Ovum 2002). Leading examples include the contracts between T-Systems and Deutsche Telekom and between Siemens Business Services and Siemens (Table 6.3). Among the top 10 suppliers in Germany, seven have a substantial share of their IT outsourcing revenue from the parent company. Nevertheless, excluding captive outsourcing from the data still leaves T-Systems in a dominant position, with IT outsourcing revenue almost double that of IBM.

Second, the client market for IT outsourcing differs. In Germany the manufacturing sector dominates and in the UK the public sector. The strong role of manufacturing – 45 per cent of the non-captive outsourcing market (PAC 2003, Figure 3.1) compared to 21 per cent in the UK (Ovum 2002, pp. 144–6) – reflects its commanding position in the German economy. More surprising is the relatively small size of the financial services

Table 6.2   The Western European IT outsourcing market, 2000–02 with forecast for 2003–06 (millions of euros)

	2000	2001	2002	2003	2004	2005	2006	Average annual growth rates		
								00–01 (%)	01–02(%)	02–06(%)
Austria	335	365	391	427	478	543	620	9.0	7.1	12.2
Belgium	640	775	870	1000	1165	1338	1527	21.1	12.3	15.1
Switzerland	1079	1113	1613	1774	1979	2201	2440	3.2	44.9	10.9
Germany	10736	12562	13454	15327	17545	20055	22802	17.0	7.1	14.1
Spain	1058	1232	1380	1603	1863	2173	2541	16.5	12.0	16.5
France	4394	5225	6044	7059	8272	9678	11333	18.9	15.7	17.0
Italy	2810	3216	3506	3903	4459	5126	5926	14.4	9.0	14.0
Netherlands	1730	2100	2600	3105	3525	3980	4485	21.4	23.8	14.6
Nordic countries	3800	4310	4812	5425	6135	6956	7905	13.4	11.6	13.2
Portugal	255	310	362	425	507	591	680	21.6	16.8	17.1
UK	12542	15276	17367	19661	22528	25425	28575	21.8	13.7	13.3
Others	452	524	589	669	766	890	1030	15.9	12.4	15.0
Total	**39831**	**47007**	**52988**	**60377**	**69222**	**78955**	**89862**	**18.0**	**12.7**	**14.1**

Note: The IT outsourcing market refers to 'outsourcing and processing services', which include takeover of data centres, applications, desktops, networks, as well as traditional processing.

Source: PAC (2003, p. 3).

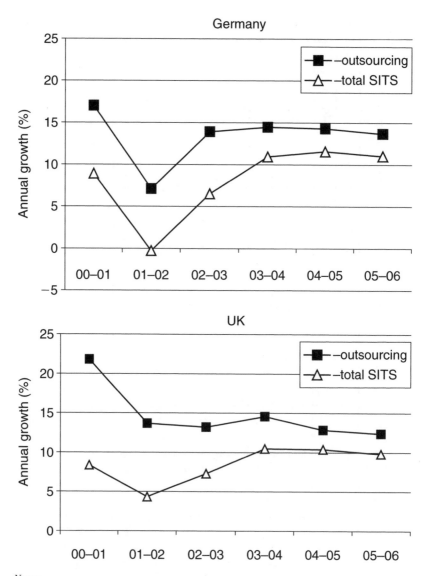

Notes:
SITS refers to the software and IT industry.
Data for 2003–06 are projected estimates and therefore ought to be interpreted with caution.

Source: PAC (2002; own calculations).

*Figure 6.1   Annual growth of IT outsourcing and total software and
              IT services industry in Germany and the UK, 2000–06*

Table 6.3   *'Captive outsourcing' among top computer services firms in the German IT outsourcing market, 2002*

Rank		Total revenue (€m)	Captive revenue (%)	Rank in non-captive market
1	T-Systems	3 920	58.4	1
2	SBS	1 439	60.8	3
3	IBM	950	–	2
4	Lufthansa Systems	578	77.0	9
5	EDS	500	–	4
6	HP	345	–	5
7	BASF IT-Services	340	98.8	–
8	Datev	310	85.5	19
9	TKIS	310	45.2	7
10	Vodaphone Information Systems	238	76.5	15
Total market		**12 920**	**32.7**	

Source: adapted from PAC (2003, Figures 3.6–3.8; own calculations).

market in the UK compared to Germany – 15 per cent and 18 per cent, respectively. In fact, the bulk of deals in the UK derive from the public sector – 33 per cent from central and local government combined – compared with a share of just 8 per cent of the (non-captive) IT outsourcing market in Germany.

But on the supplier side both countries reveal a similarly high level of concentration. While it is difficult to construct accurate measures of market shares, estimates for 2002 in Germany and 2001 in the UK suggest the top ranking firm controlled 30 per cent and 26 per cent, respectively, of the IT outsourcing market and the respective shares for the top five firms were 57 per cent and 61 per cent (Ovum 2002, Figure H3.1; PAC 2003, Figures 3.7, 3.8; own calculations). The leading US-owned IT firms have enjoyed a strong foothold in both markets; four firms – EDS, IBM, CSC and Schlumberger/Sema (US–French) – appeared in the top 20 in both Germany (2002) and the UK (2001). There have been fewer crossovers of UK and German-owned firms to date, with just one leading German firm in the UK top 20 (Siemens Business Services) and no UK firms in the German top 20. The predominance of multinationals has also been associated with a growing number of 'megadeals' (contracts valued at more than €100 million), although these emerged later in Germany than in the UK.

Deliberative Institutions

All 13 IT outsourcing contracts investigated involved staff transfer, ranging from 20 to approximately 12 000 employees (Table 6.4). Our interviews revealed a consensus among managers in the IT firms that the transfer of IT staff from the client organization was essential for acquiring firm-specific knowledge of IT systems and business processes; as one manager put it, 'losing key people in those transformation processes is the death of outsourcing' (general interview, IT firm, Germany). Client and supplier managers in both countries identified positive opportunities for transferees, especially for those seeking specialist, or international, career development. However, reflecting other studies (George 2003), managers were also conscious of the risk and uncertainty associated with short-term contracts (between 5 and 10 years) and successive changes in employer.

Processes of managing staff transfer differed substantially between the two countries. In the six German contracts, extensive negotiations preceded outsourcing, typically involving six months of communications with works councils. No substantive differences in policy were found among the three IT firms by country of ownership, reflecting a dominant 'host country effect' over potential variation caused by 'corporate' or 'home country effects' (Ferner and Quintanilla 1998). As found in other research on malleability to trade union pressures (Edwards and Ferner 2002), the selected US and UK-owned IT firms responded to works councils' pressures, in common with the German-owned firm. The result was a stronger determination among managers in Germany to win the 'hearts and minds' of their IT workers and to design a process of restructuring IT services favourable to staff adjustment. In four of the six German contracts (Table 6.4), this involved transferring staff to a newly created organization – a joint venture between client and supplier.

> The first reason [for joint ventures] is human resources . . . because if you are creating a joint venture you can transfer those resources very smoothly. (general interview, IT firm, Germany)

> The German [client] firms say, 'A joint venture – it's better for our employees'. . . . The employees feel that they are part of [the client organization]'. (Contract 10, IT firm, Germany)

The evidence suggests that negotiations with works councils could speed up or slow down outsourcing in Germany. At one of the foreign IT firms, in one contract (Contract 10) the works council of the client organization had been actively in favour of it winning the IT deal and had facilitated the process. But in another contract (Contract 13), the works council had favoured a German IT firm and this had slowed the process significantly.

Table 6.4 Staff transfer and organizational form in 13 contracts

Contract no.	Country of operation	Foreign or domestic IT firm*	Number of staff transferred	Organizational form	Additional details
1	UK	F	1 600	Direct outsourcing	IT outsourced in 1998. Client cancelled the 10-year contract in 2003 and insourced IT.
2	UK	F	2 300	Direct outsourcing	IT outsourcing contract began in 1994, but staff transferred in two stages, 1994 and 1996.
3	UK	D	30	Direct outsourcing	IT outsourced in 2000.
4	UK	F	420	Direct outsourcing	IT outsourced in 2001. Post-transfer downsizing of staff to 360 by IT firm.
5	UK	F	1 600	Direct outsourcing	IT outsourced in 2000.
6	UK	F	1 190	Direct outsourcing	IT outsourced in 1996–97 and a new 12-year deal was agreed in 2000.
7	UK	D	20	Direct outsourcing	IT outsourced in 1995 and contract renewed in 2000.
8	DE	F	28	Joint venture	IT outsourced in 1997–98 to a joint venture with a German-owned IT firm. IT firm acquired by a foreign multinational in 2000 and its share in the joint venture increased from 33% to 52% (2003); expected to increase to 100%.
9	DE	F	20–30	Joint venture	IT outsourced to a joint venture in 1998. IT firm increased share to 100%.
10	DE	F	280	Joint venture	IT outsourced in 1995 to a 25%–75% joint venture between the client and the IT firm. The IT firm then grew its share to 100%.

Table 6.4 (continued)

Contract no.	Country of operation	Foreign or domestic IT firm*	Number of staff transferred	Organizational form	Additional details
11	DE	D	Approx. 12 000	Captive market	Client established a 100% owned affiliate. In 2001 the affiliate acquired another IT firm, providing services to third party clients.
12	DE	D	55	Joint venture	IT outsourced in 1996 to a joint venture with an IT firm, which acquired 100% of shares in 2000. In 2001 it was bought by another German IT firm.
13	DE	F	900	Captive market/ direct outsourcing	IT outsourced in the mid-1990s to a 100% owned affiliate. IT then outsourced from the affiliate to the foreign-owned IT firm.

Note: * Information on country of ownership is limited to foreign (F) and domestic (D) to preserve confidentiality.

All three IT firms investigated in Germany shared similar intentions with regard to the longer-term organizational form, using joint ventures only as transitional arrangements.

In all seven UK contracts, management of staff transfer involved only minimal consultation and sharing of information with worker representatives. In six contracts, there was an immediate switch from complete in-house provision to externalized provision; we found no evidence that consultations with staff influenced the form of IT outsourcing arrangement (aside from the general need to respect legal regulations on employment protection at the point of transfer). In a seventh, Contract 2, IT staff staged a one-day strike prior to the transfer and refused to work overtime. Also, the staff association organized an intensive lobbying campaign and planted questions in Parliament. As a result, staff transfer was organized in two stages, with 1200 transferring in 1994 and 1100 in 1996. Despite these efforts, the IT firm in Contract 2 still experienced problems of alienation and resistance among transferees – a number of whom were said to be 'very resistant' and wanting 'to turn the clock back' (Contract 2, IT firm, UK). In part, problems at Contract 2 reflected the specific issues of transferring from a public to a private sector employer – an issue we did not investigate in Germany given the sample. But our evidence also suggests that industrial relations problems were more pervasive in the UK than in Germany. Managers in four of the seven UK contracts (Contracts 1, 2, 6, 7) reported experiencing staff resistance to transfer, yet in the absence of institutionalized mechanisms for 'employee voice' (Freeman and Medoff 1985) such resistance had little impact on the strategy of outsourcing. As one client manager put it, 'Suddenly they were told, '[the client] is no longer interested in IT' – that means I am a second-class citizen, they don't want me anymore. . . . It was probably not well handled at the time, because it was seen very much as a cost issue' (Contract 7, client, UK). In Germany, we uncovered only one example of staff resistance; in Contract 12 some senior IT workers had exercised their right to refuse to transfer and remained employees of the client organization providing freelance IT services. Further research involving data from employees and/or works councils representatives might shed further light on the issue of staff resistance to IT outsourcing.

We also found country differences in the extent to which client organization managers were satisfied with the partner IT firm's human resource policies (especially for recruitment and training), with greater evidence of dissatisfaction in the UK than in Germany. Problems cited by UK client managers included a deterioration of services quality caused by new recruitment practices in the IT firm (Contract 2) and the loss of high-skilled IT staff because the IT firm creamed off the 'top talent' to work on other contracts (Contract 1). By contrast, the four joint ventures and the one example of captive outsourcing in Germany seemed to provide client managers with a more effective arrangement to coordinate the skill-mix of IT workers. However, when joint ventures were subsequently bought out by the IT firm, German client managers anticipated problems, as the IT firm siphoned off the best staff to other contracts. At Contract 8, where the joint venture still prevailed at the time of research, the client manager told us:

> I have been in the outsourcing business [with a multinational IT firm]. We bought a whole group of customers, set up a new company [joint venture] . . . After a year or two or three, we took 100 per cent of the good consulting people and . . . spread them all over the world. And the customer never saw the good people again. Gone. And that's what we don't want to see happen over here. (Contract 8, client, Germany)

Client Expertise

Our data reveal differences across contracts in the relative expertise of personnel in client organizations in managing the technological and con-tractual requirements of IT outsourcing. We have constructed measures of high, medium and low for three features of client expertise for all 13 contracts (Table 6.5) based on documentary information and the responses of project managers in the IT firms to questions such as: 'to what extent does your customer understand the technology you are dealing with?'; 'how adept is the client at monitoring IT services delivery?'; and 'how effective is the benchmarking of prices utilized by the client organization?'

In all contracts except one (Contract 6), managers argued that a client's knowledge of advances in IT was essential to monitor IT delivery and align IT strategy with business needs. Nevertheless, our data suggest varying levels of technical knowledge among clients. In the UK, the disparity of technical expertise among clients was more apparent than in Germany where levels tended to be universally high. Considering two UK clients, Contracts 3 and 7, which outsourced to the same IT firm, we found high and low technical knowledge, respectively. For Contract 3, this reflected a policy of retaining a strong internal team of 17 technical staff (including six senior managers) to advise on IT strategy and to monitor IT services delivery. At Contract 7, low technical knowledge led to misunderstandings concerning the technical specifications in the contract – for example, regarding the expected level of service, innovation and software develop-ment – and generated unanticipated additional costs for the client. Also, in some contracts where UK client managers did not have 'high' technical knowledge, this appeared inadvertently to have fostered opportunistic behaviour by the partner IT firm. At Contract 1, the IT firm project

Table 6.5 Comparing features of client expertise in 13 contracts

Contract no.	Country of operation	Technical knowledge of client	Contract expertise of client	Exercise of market discipline
1	UK	Medium	Low	Low
2	UK	Medium	Low	Low
3	UK	High	High	High
4	UK	High	Medium	n.a.
5	UK	Medium	Low	Low
6	UK	Medium	Medium	Low
7	UK	Low	Medium	High
8	DE	High	High	Low
9	DE	High	High	Medium
10	DE	High	High	n.a.
11	DE	High	Medium	High
12	DE	Medium	High	High
13	DE	High	Medium	n.a.

Note: The measures high, medium and low are qualitative indicators based on documentary information and the responses of project managers in IT firms to open-ended questions (see main text).

manager was confident about passing on to the client only a fraction of the benefits of efficiency improvements, by being able to 'improve things behind the scene without them knowing' (IT firm, UK).

Evidence of strong technical knowledge among German clients appeared to provide a basis for greater cooperation between partners (Burchell and Wilkinson 1997) and for the effective delivery of IT services. A significant contributing feature to client expertise was the adoption of joint ventures, since the client organization retained strong investment and influence over technological developments. Another factor (also found in some UK contracts) was the retention of in-house IT expertise, although numbers of staff were lower in contracts where joint ventures still prevailed, compared to those where they had been bought out by the IT firm. For example, retained staff numbered four in the Contract 8 client and 70 and 55, respectively, in Contracts 10 and 12. Overall, project managers from IT firms in Germany were far more likely than those in the UK to perceive that client project managers had a high level of technical knowledge. The following quote is illustrative:

Everybody [on the client side] is specialized on a certain part of the outsourcing contract. There is one guy specializing on [mainframe services] . . . He for sure is the very best of all. . . . So everything that deals with mainframes will be approved or requested by this guy. . . . And, like this guy, there are specialists for mid-range systems, for client architecture, for telecommunications, whatever. . . . They are very, very good (Contract 9, IT supplier, Germany).

Our data also suggest variability in client expertise in designing and managing contracts. Three of the 13 client organizations – all in the UK – registered 'low' contract expertise. In Contract 1, following problems during the initial years of the contract the client came under pressure from its headquarters office (it is a UK subsidiary of a US multinational) to impose stricter control over contract respecifications. Its failure to do so led to a subsequent decision to cancel the 10-year contract after just five years. Contracts 2 and 5 both involve public sector clients where low contract expertise reflected the recent opening of public sector markets and the slow development of contracting skills among civil servants (Gershon 1999). In Contract 2, government regulatory bodies investigated the expertise of client managers in response to concerns about the doubling of the initial value of the 10-year contract. A National Audit Office investigation found problems with certain skills (e.g. client project managers' knowledge of pricing mechanisms), evidence of insufficient resources and expertise in the contract assessment units, and a lack of foresight in replenishing skills.

Most examples of 'high' contract expertise were in Germany. In Contract 8, the client was perceived to have 'a strong outsourcing strategy' (IT firm, Germany) in which client senior project managers had negotiated full access to the IT firm's database of performance statistics. In Contract 10, strong contract expertise was perceived to have facilitated a relatively cooperative approach in dealing with conflicts between the two partners, especially over a lengthy renegotiation of the contract. And in Contract 12 the client had developed an innovative contract enabling high flexibility for cost reduction or service changes, with limited risk of the IT firm charging for contract respecification.

A third feature of client expertise concerns the exercise of 'market discipline' over IT firms' performance. Examples of practices include the external benchmarking of prices (e.g. for hardware, software licences and 'function points' – a measure of the size of computer applications and the projects that build them) and the use of multi- as opposed to single-supplier contracts. In five contracts (four in the UK), we found little evidence of benchmarking (either in scope or effectiveness). In Contract 2, client managers could only benchmark half of all outsourced IT services due to difficulties in obtaining confidential information from competing IT firms and in constructing comparable measures. In Contract 6, under the guise of an unusually strong partnership approach between the UK client and IT firm, the client had delegated responsibility for benchmarking to the IT firm. And in Contract 8, the German client faced increasing pressure from head office to strengthen market discipline, in response to the IT firm increasing its shares of the joint venture. Both Germany and the UK display two contracts where clients combined an effective benchmarking strategy with a multi-supplier contracting arrangement, contributing to 'high' market discipline. It is notable that in all four contracts the IT firm was not foreign-owned. This appeared to grant greater power to the client:

> We had a lot of switches [of suppliers] in the past because of savings we could realize . . . We want to pay as little as possible. . . . We are also allowed to do benchmarking . . . and if the benchmarking proves that we are right the supplier has to pay. (Contract 12, client, Germany)

> [The client chief executive] knows the people that run IBM and EDS. They have dinner and they say, 'We can do that for you'. They've got contacts and because these guys are bigger than we are, then they will tend to get in. (Contract 7, IT firm, UK)

Also, in Contract 12, which spanned several countries the client signed with multiple suppliers in response to the limited global presence of the German-owned IT firm, thus increasing pressure on the IT firm to perform well. By contrast, in contracts involving a single-supplier contract, client managers were more at risk of becoming locked into an outsourcing arrangement with the supplier. In both countries, interviewees recognized a high risk of lock-in, reflecting the strongly concentrated base of suppliers and the perceived difficulties of transferring idiosyncratic assets (and staff) to a new IT supplier.

DISCUSSION

This section assesses our exploratory results on the institutional effects on the market for IT outsourcing and identifies contributions to theory development. Possibilities for future research are highlighted where appropriate.

A first issue was the extent to which expansion of IT outsourcing markets in Germany and the UK was comparable, or, as suggested by literature on Germany's 'service gap', different. Our findings show that although the IT outsourcing market developed later in Germany than in the UK, both countries enjoyed remarkably high growth during the late 1990s and early 2000s. Such evidence runs counter to the idea that since KIBS markets depend on radical restructuring and deverticalization of organizations (as they switch from in-house to external business services provision) then only a liberal market economy, such as the UK, provides a good institutional fit.

But does this suggest convergence of diverse political economies on a uniform model of market expansion? Our analysis of industry-level data suggests not. The conditions underpinning market growth were country-specific. In the UK, government emphasis on public–private partnerships (IPPR 2001) fuelled market growth by developing a strong public sector demand for IT services. In Germany, two conditions were pertinent: captive markets established a stable market platform for German IT firms to sell IT services to other client organizations; and strong demand from manufacturing firms – reflecting the services-intensive nature of high-tech manufacturing found in Germany (Ietto-Gillies 2002) – increased the scope for linkages with the computer services sector. Nor, however, does our evidence support the notion that markets for IT outsourcing have developed in country-specific equilibria, defined by distinctive employer strategies within diverse institutional settings. Here, we depart from a tendency within the varieties of capitalism approach (Hall and Soskice 2001) to stress institutional resiliency, static diversity of market forms and limited degrees of freedom for actors to exercise choice within relatively stable, path-dependent institutional configurations (Blyth 2003; Thelen and van Wijnbergen 2003). A notable feature of market expansion in both countries has been the increasing concentration of market share among US multinational IT firms and their control over the market for 'mega-deals'. The dominant role played by MNCs, coupled with the fast pace of technological change in this sector, established a basis for common systemic trends (Strange 1997), albeit, as we found in the UK and German IT outsourcing markets, mediated by institutional and organizational factors.

One important institutional factor was the role played by deliberative institutions in shaping the pace and trajectory of growth of the IT outsourcing market. Confirming the more general argument (Sabel 1994), because the transaction embodying each IT outsourcing contract involved distributive dilemmas, the relatively strong deliberative institutions in Germany in fact played a facilitating, rather than a hindering, role. Our evidence shows that two distributive dilemmas were central to the expansion of the IT outsourcing market – one related to the transfer of staff from the client organization to an IT firm and the other to uncertainty over the distribution of performance gains between client and IT firm. Joint gains to the transferring IT workforce, the IT firm and the client organization were possible (improved career opportunities, immediate acquisition of tacit knowledge of client business practices, and continued access to ex-employees' skill-sets, respectively), but the distribution of gains depended on the strategy adopted.

In Germany, 'specific conceptions of distributive justice' (Hall and Soskice 2001, p. 12) were established through successive negotiations between client employers, IT firm employers and works councils. With strong negotiating rights (Slomp 1995), works councils could pressure IT firms to establish joint ventures with clients, thereby reassuring transferring IT workers of the client's ongoing commitment to their employment security. Moreover, deliberation also assisted the client organization since the joint venture provided an organizational form that facilitated the coordination – between client and IT firm – of human resource practices, especially regarding the management of skill-mix among workers charged with delivering IT services. In the UK, with relatively weak deliberative institutions, IT outsourcing involved minimal consultation with staff and an immediate switch from in-house to externalized provision. The distribution of gains appeared less favourable from the perspective of both the workforce and the client organization. Our data revealed examples of industrial relations problems. Also, client managers were more likely than their counterparts in Germany to complain of poor services quality caused by IT firm practices of replacing more able staff with less experienced new recruits (so-called 'cream-skimming').

Our analysis of the country-specific character to the role of deliberative institutions in shaping the outcomes of these distributive dilemmas is complicated by the intersection between employment relations and changing organizational forms in shaping the market for IT outsourcing. Theoretical analysis on changing organizational forms has been impeded by its disconnection with the study of changes in work and employment (Barley and Kunda 2001). Our argument is that where new markets involve the spread of inter-organizational relations, this impacts on the authority relations under which employment is organized, opening up new arenas for conflict and cooperation (Grimshaw et al. 2004; Rubery et al. 2002). The role played by German works councils ought not therefore, simply be interpreted as symptomatic of an institutional enabling role in the economy. Rather, particular actors (works council representatives and client organization managers) responded to the new employment challenges posed by a shift to inter-organizational relations and were able to operationalize Germany's institutional capacity for strategic action. Much of what we found is thus contingent upon the particular characteristics of market growth in the IT outsourcing sector, involving close client–supplier relations.

But, returning to the issue of convergence and divergence, our findings suggest that country differences in the trajectory of market expansion were not stable. All joint ventures investigated in Germany were transitional forms. In each contract, after a short period the client's ownership share was bought out (or expected to be bought out) by the partner multinational

IT firm. Medium-term convergence with organizational forms (i.e. direct client–supplier relations) found in liberal market economies like the UK was accompanied by expectations among German client managers of heightened tensions surrounding the deployment of IT workers providing services. Such radical change in the 'fit' between organizational form and institutions in the German IT outsourcing market occurred through a slow process of diffusion of new business practices, largely shaped by the 'challenger rules' of powerful multinational IT firms (Djelic and Quack 2003), which called into question Germany's institutionally specific practice of establishing joint ventures. Others refer to this type of rupture as 'displacement through invasion' (Streeck and Thelen 2005), whereby new models emerge not through explicit amendment of existing institutions but 'shifts in the relative salience of different institutional arrangements' (ibid., p. 33). Thus while there was no change in the rules governing deliberative institutions, the German 'solution' to the distributive dilemmas posed by IT outsourcing – the joint venture – was ultimately replaced by a universal organizational model adopted by multinational IT firms in liberal and coordinated market economies alike.

A focus on institutional effects also contributes to theoretical development regarding the types of capabilities that enable client organizations to shape the quality of KIBS provision and benefit from improved performance through relations with an external KIBS provider firm (Earl 1996; Gallouj 1997). Effective inter-organizational linkages are seen to be fundamental to realizing the potential for KIBS firms to improve competitiveness in other sectors through providing intermediate inputs (Peneder et al. 2003). In a review of research, Miles (2003) suggests that the potential for knowledge and technology spillovers from a KIBS firm to a client organization increases where there is close and continuous interaction, strong trusting and equitable relations ('sparring relations' in Tordoir's (1995) terminology), geographical proximity and client prior knowledge (ibid., pp. 37–50).

While sensitive to the characteristics and distinctiveness of the KIBS sector, such studies are limited in scope by their lack of attention to the way country institutions configure client capabilities. Our research shows, for example, that German client managers tended to be perceived as having greater expertise than UK clients, reflecting, in part, differential institutional effects resulting from legal rules and technical standards in each country (Arrighetti et al. 1997; Lane 1997). However, these institutional effects were not deterministic; they interacted with other conditions associated with the selected KIBS market, IT outsourcing. First, some client organizations pursued the practice of retaining an in-house IT workforce. This organizational practice enhanced a client's 'absorptive capacity'

(Cohen and Levinthal 1990) by sustaining the level of technical knowledge and explained much of the intra-country variation in expertise among client managers. Second, in the UK weak contract expertise was more likely among public sector clients than private sector clients (no public sector contracts were explored in Germany reflecting its small share of the client market). In two of three examples of weak contract expertise in the UK, this was partly explained by the limited experience of public sector managers in administering private sector contracts (Grimshaw et al. 2002; HM Treasury 2003). Third, IT outsourcing contracts differed according to the degree of market discipline exercised by the client organization. In both countries, clients enjoyed greater bargaining power when contracting with a domestically owned IT firm. However, there were more examples of ineffective benchmarking and reliance on single-supplier contracts in the UK than in Germany, suggesting that weak technical and contract-based expertise could restrain clients' capacity to seek competing bids or exploit multi-supplier contracts. As one of the German client managers explained, 'If you have the competence, you can also look for competitors. . . . It's the power to get every price' (Contract 9, client, Germany).

Thus, our study of a single market for KIBS responds to the need to explore institutional effects on the capacity of client organizations to shape, and benefit from, KIBS provision. At the same time, to a greater extent than found in other cross-national studies on contracting (Burchell and Wilkinson 1997; Lane 1997), it also highlights the non-deterministic effects of a country's institutions. Institutional effects interact with sectoral characteristics, involving organizational practices (e.g. retention of IT staff post-outsourcing), the composition of the client market (e.g. public vs. private sector clients) and relative bargaining power of clients and providers. The outcome is variability in the degree of client expertise within both countries, yet around a higher level in Germany than in the UK.

CONCLUSION

Our research suggests that the institutional context in Germany and the UK plays a significant role in shaping the development and form of a fast-growing area of knowledge intensive business services, IT outsourcing. While the liberal market economy of the UK might be considered more conducive to contracting and outsourcing, our study found that the coordinated market economy of Germany in fact facilitated comparable growth in the market for IT outsourcing. Instead, institutional effects were apparent in the distinctive form and consequences of contracting arrangements between client organizations and IT firms.

Our findings support the results of other studies on the institutional effects on contracting arrangements, but our focus on KIBS highlights the importance of the sectoral context. A first characteristic is the important role of knowledge assets – in the form of intellectual and human capital – in driving expansion in KIBS sectors (Starbuck 1992; Swart and Kinnie 2003; Teece 1998). Fast growth in the IT outsourcing market means that IT firms need to expand knowledge assets and competences rapidly. However, such assets are not readily available in a given market: non-separability of IT and production technologies (Jonscher 1994) means that tacit knowledge of client business processes is a valuable component of IT knowledge; and pressure to fulfil large contracts with client organizations means continuity of services provision is required from day one of the contract. Together these conditions explain the apparent pervasiveness of staff transfers with IT outsourcing. But viewed as an institution for trading knowledge assets, the market for staff transfers has certain peculiarities. Teece (1998, pp. 67–9) identifies problems of conflict over ownership rights, difficulties in valuing knowledge assets and in establishing the unit of consumption. Our study identifies a further problem – what we refer to as the distributional dilemma of staff transfers. Transfer of substantial numbers of highly skilled IT workers from client organization to IT firm presented potential performance gains and losses for the client, the IT firm and the transferring workers. Our data from Germany show that institutions could facilitate trading of knowledge assets through deliberative processes (works councils and code-termination) that reduced the risks of the distributive dilemma. In the UK, lack of deliberation did not diminish the trading of IT workers, but did underpin subsequent performance losses experienced by some client organizations – especially associated with problems in coordinating services quality with human resource policy in the partner IT firms – and fostered worker resistance to transfer in the absence of institutionalized mechanisms for voice. Observed institutional effects were thus closely associated with the important role of knowledge assets in the market for IT outsourcing.

A second characteristic is the strong internationalization in many KIBS sectors (Miozzo and Miles 2002). Attention to the characteristics of the supply base, especially distinguishing market share and the parent country of multinationals, further refines our analysis of institutional effects on the form of contracting and extends previous cross-national studies of differences in contracting (e.g. Arrighetti et al. 1997). Our data suggest two findings, as well as possible avenues for further research. One finding was that when client organizations contracted with an IT firm that was not US-owned, they were able to exercise greater market discipline – reflecting the fact that US-owned IT firms enjoy leading market positions and stronger bargaining power. Power relations between client and IT firm thus mediated institutional effects. A second finding was that multinational IT firms played a dynamic role in establishing 'best practice' in IT outsourcing markets, contributing to a convergence in organizational forms. What is uncertain is the degree to which country institutions – specifically mechanisms for employee voice and bodies to diffuse contracting expertise – will adapt to converging practices of multinational IT firms in shaping this fast-growing KIBS market. Further research could explore the strength of MNCs' 'challenger rules' (Djelic and Quack 2003) in KIBS markets in standardizing business and HR practices, and the extent to which these are mediated by host country institutions (Edwards and Ferner 2002) and 'network effects' arising from pressures to conform to the needs of client organizations and affected workers.

ACKNOWLEDGEMENT

The authors are grateful for the financial support of the Anglo-German Foundation (grant no. 1360), research assistance from Paulina Ramirez and the contributions of Matthias Knuth and Thorsten Kalina. Mark Lehrer generously provided helpful comments on previous drafts.

REFERENCES

Ackroyd, S. and S. Procter (1998), 'British manufacturing organization and workplace industrial relations: some attributes of the new flexible firm', *British Journal of Industrial Relations*, **36** (2), 163–83.

Antonelli, C. (1998), 'Localised technological change, new information technology and the knowledge-based economy', *Journal of Evolutionary Economics*, **8** (2), 177–98.

Arrighetti, A., R. Bachmann and S. Deakin (1997), 'Contract law, social norms and inter-firm cooperation', *Cambridge Journal of Economics*, **21** (2), 171–95.

Atiyah, P. (1995), *An Introduction to the Law of Contract*, Oxford: Oxford University Press.

Barley, S. and G. Kunda (2001), 'Bringing work back in', *Organization Science*, **12** (1), 76–95.

Blyth, M. (2003), 'Same as it never was: temporality and typology in the varieties of capitalism', *Comparative European Politics*, **1** (2), 215–25.

Bosch, G. (2001), 'Germany: a "service gap"?,' in J.E. Dølvik (ed.), *At Your Service? Comparative Perspectives on Employment and Labour Relations in the European Private Sector Services*, Brussels: P.I.E. – Peter Lang, pp. 53–101.

Burchell, B. and F. Wilkinson (1997), 'Trust, business relationships and the contractual environment', *Cambridge Journal of Economics*, **21** (2), 217–37.

Casper, S. (2001), 'The legal framework for corporate governance: the influence of contract law on company strategies in Germany and the US', in P. Hall and

D. Soskice (eds), *Varieties of Capitalism: The Institutional Foundations of Comparative Advantage*, Oxford: Oxford University Press, pp. 387–416.

Coe, N. (2000), 'The externalisation of producer services debate: the UK computer services sector', *Services Industries Journal*, **20** (2), 64–81.

Cohen, W. and D. Levinthal (1990), 'Absorptive capacity: a new perspective on learning and innovation', *Administrative Science Quarterly*, **35** (1), 128–52.

Colling, T. (2000), 'Personnel management in the extended organization', in S. Bach and K. Sisson (eds), *Personnel Management: A Comprehensive Guide to Theory and Practice*, Oxford: Blackwell, pp. 70–90.

Coombs, R., M. Harvey and B. Tether (2003), 'Analysing distributed processes of provision and innovation', *Industrial and Corporate Change*, **12** (6), 1125–55.

Crouch, C. and W. Streeck (eds) (1997), *Political Economy of Modern Capitalism: Mapping Convergence and Diversity*, London: Sage.

Cully, M., S. Woodland, A. O'Reilly and G. Dix (1999), *Britain at Work*, London: Routledge.

Culpepper, P. and D. Finegold (eds) (1999), *The German Skills Machine in Comparative Perspective*, Oxford: Berghahn Books.

Daniels, P. and F. Moulaert (1991), *The Changing Geography of Advanced Producer Services: Theoretical and Empirical Perspectives*, London: Belhaven Press.

Djelic, M.L. and S. Quack (2003), 'Introduction: governing globalization – bringing institutions back in', in M.L. Djelic and S. Quack (eds), *Globalization and Institutions. Redefining the Rules of the Economic Game*, Cheltenham, UK and Northampton, MA, USA: Edward Elgar, pp. 1–14.

Earl, M. (1996), 'The risks of outsourcing IT', *Sloan Management Review*, **37** (3), 26–32.

European Commission (EC) (1997), *Industrial Competitiveness and Business Services: Report to the Industry Council*, Brussels: European Commission.

European Commission (EC) (2003), 'The competitiveness of business-related services and their contribution to the performance of European enterprises', communication from the Commission to the Council, the European Parliament, the European Economic and Social Committee and the Committee of the Regions, COM (3002) 747, Brussels: European Commission.

Edwards, T. and A. Ferner (2002), 'The renewed "American challenge": a review of employment practice in US multinationals', *Industrial Relations Journal*, **33** (2), 94–111.

European Industrial Relations Observatory (EIRO) (2001), 'Industrial relations consequences of mergers and takeovers: the case of Germany', accessed 10 October, 2003, at www.eiro.eurofound.eu.int/2001.

Eisenhardt, K. (1989), 'Building theories from case study research', *Academy of Management Review*, **14** (4), 532–50.

Ferner, A. and J. Quintanilla (1998), 'Multinationals, national business systems and HRM: the enduring influence of national identity or a process of "Anglo-Saxonization"', *International Journal of Human Resource Management*, **9** (4), 710–31.

Freeman, R. and J. Medoff (1985), *What do Unions do?* New York: Basic Books.

Froud, J., C. Haslam, S. Johal and K. Williams (2000), 'Shareholder value and financialization: consultancy promises, management moves', *Economy and Society*, **29** (1), 80–111.

Gallouj, C. (1997), 'Asymmetry of information and the service relationship: selection and evaluation of the service provider', *International Journal of Service Industry Management*, **8** (1), 42–64.

George, E. (2003), 'External solutions and internal problems: the effects of employment externalisation on internal workers' attitudes', *Organization Science*, **14** (4), 386–402.

Gershon, P. (1999), 'Review of civil procurement in central government', accessed 18 October, 2003, at www.ogc.gov.uk/gershon/ pgfinalr.htm.

Goode, R. (1995), *Commercial Law*, Harmondsworth: Penguin.

Grimshaw, D., S. Vincent and H. Willmott (2002), 'Going privately: partnership and outsourcing of public sector services', *Public Administration*, **80** (3), 475–502.

Grimshaw, D., J. Rubery, M. Marchington and H. Willmott (2004), 'Bringing employment back in: a critique of current theorizing of new organizational forms', paper presented at the Work, Employment and Society Conference, University of Manchester (September).

Hall, P. and D. Soskice (2001), 'An introduction to varieties of capitalism', in P. Hall and D. Soskice (eds), *Varieties of Capitalism: The Institutional Foundations of Comparative Advantage*, Oxford: Oxford University Press.

Hislop, D. (2002), 'The client role in consultancy relations during the appropriation of technological innovations', *Research Policy*, **31** (5), 657–71.

HM Treasury (2003), *PFI: Meeting the Investment Challenge*, London: Stationery Office.

Hollingsworth, J. R., P. Schmitter and W. Streeck (eds) (1994), *Governing Capitalist Economies: Performance and Control of Economic Sectors*, Oxford: Oxford University Press.

Hölmstrom, B. (1985), 'The provision of services in a market economy', in R. Inman (ed.), *Managing the Service Economy: Prospects and Problems*, Cambridge: Cambridge University Press, pp. 183–213.

Hyman, R. (2001), *Understanding European Trade Unionism: Between Market, Class and Society*, London: Sage.

Ietto-Gillies, G. (2002), 'Internationalization and the demarcation between services and manufactures: a theoretical and empirical analysis', in M. Miozzo and I. Miles (eds), *Internationalization, Technology and Services*, Cheltenham, UK and Northampton, MA, USA: Edward Elgar, pp. 33–56.

IPPR (Institute for Public Policy Research) (2001), *Building Better Partnerships: The Final Report of the Commission on Public Private Partnerships*, London: IPPR.

Jackson, G., M. Höpner and A. Kurdelbusch (2004), 'Corporate governance and employees in Germany: changing linkages, complementarities and tensions', in H. Gospel and A. Pendleton (eds), *Corporate Governance and Labour Management: An International Comparison*, Oxford: Oxford University Press, pp. 84–121.

Jonscher, C. (1994), 'An economic study of the information technology revolution', T. Allen and M. Scott Morton (eds), in *Information Technology and the Corporation of the 1990s*, Oxford: Oxford University Press, pp. 5–42.

Lacity, M.C. and L. Willcocks (2001), *Global Information Technology Outsourcing: In Search of Business Advantage*, Chichester: Wiley.

Lane, C. (1989), *Management and Labour in Europe: The Industrial Enterprise in Germany, Britain and France*, Aldershot, UK and Brookfield, US: Edward Elgar.

Lane, C. (1997), 'The social regulation of inter-firm relations in Britain and Germany: market rules, legal norms and technical standards', *Cambridge Journal of Economics*, **21** (2), 197–215.

Lane, C. and R. Bachmann (1996), 'The social constitution of trust: supplier relations in Britain and Germany', *Organization Studies*, **17** (3), 365–95.

Lane, C. and R. Bachmann (1997), 'Co-operation in inter-firm relations in Britain and Germany: the role of social institutions', *British Journal of Sociology*, **48** (2), 226–54.

Lehrer, M. (2005), 'Two types of organizational modularity: SAP ERP product architecture and the German tipping point in the make/buy decision for IT services', in M. Miozzo and D. Grimshaw (eds), *Institutions of the New Economy: Knowledge Intensive Business Services and Changing Organizational Forms*, Cheltenham,UK and Northampton, MA, USA: Edward Elgar.

Lehrer, M. and O. Darbishire (1997), 'The performance of economic institutions in a dynamic environment: air transport and telecommunications in Germany and Britain', Wissenschaftszentrum für Sozialforschung discussion paper, FS I 97 – 301, Berlin.

Marchington, M., D. Grimshaw, J. Rubery and H. Willmott, (eds) (2005), *Fragmenting Work: Blurring Organizational Boundaries and Disordering Hierarchies*, Oxford: Oxford University Press.

Marsden, D. (1999), *A Theory of Employment Systems: Micro-foundations of Societal Diversity*, Oxford: Oxford University Press.

McGivern, C. (1983), 'Some facets of the relationship between consultants and clients in organizations', *Journal of Management Studies*, **20** (3), 367–86.

Meyer-Krahmer F. (2001), 'The German innovation system', in P. Larédo and P. Mustar (eds), *Research and Innovation Policies in the New Global Economy: An International Comparative Analysis*, Cheltenham, UK and Northampton, MA, USA: Edward Elgar, pp. 205–52.

Miles, I. (2001), 'Knowledge-intensive business services revisited', Nijmegen Lectures on Innovation Management, Maklu, Antwerpen-Apeldoorn.

Miles, I. (2003), 'Knowledge-intensive services' suppliers and clients: a review', mimeo, Manchester Business School: University of Manchester, January.

Miles, M.B. (1979), 'Qualitative data as an attractive nuisance: the problem of analysis', *Administrative Science Quarterly*, **24** (4), 590–601.

Miles, M. and M. Huberman (1994), *Qualitative Data Analysis: An Expanded Sourcebook*, Thousand Oaks, CA: Sage.

Miozzo, M. and I. Miles (2002), *Internationalization, Technology and Services*, Cheltenham, UK and Northampton, MA, USA: Edward Elgar.

Nelson, R. (1993), *National Innovation Systems*, New York: Oxford University Press.

Organisation for Economic Co-operation and Development (OECD) (1999), *Science, Technology and Industry Scoreboard 1999, Benchmarking Knowledge-based Economies*, Paris: OECD.

Ovum (2001), *The Holway Industry Report 2001: The Definitive Guide to UK Software and IT Services Companies and Markets*, London: Ovum.

Ovum (2002), Holway@*Ovum: Market Trends 2002*, London: Ovum.

Peneder, M., S. Kaniovski and B. Dachs (2003), 'What follows tertiarisation? Structural change and the role of knowledge-based services', *Services Industries Journal*, **23** (2) 47–66.

Pierre Audoin Consultants (PAC) (2003), *Outsourcing Program Germany: The Outsourcing and Processing Services Industry in Germany*, Munich: PAC.

Rubery, J., J. Farnshaw, M. Marchington, F.L. Cooke and S. Vincent (2002), 'Changing organisational and the employment relationship', *Journal of Management Studies*, **39** (5), 645–72.

Sabel, C. (1994), 'Learning by monitoring: the institutions of economic development', in N. Smelser and R. Swedberg (eds), *Handbook of Economic Sociology*, Princeton, NJ: Princeton University Press, pp. 137–65.

Sako, M. (1992), *Prices, Quality and Trust*, Cambridge: Cambridge University Press.

Slomp, H. (1995), 'National variations in worker participation' in A.-W. Harzing and J. Van Ruysseveldt (eds), *International Human Resource Management*, London: Sage, pp. 291–317.

Starbuck, W. H. (1992), 'Learning by knowledge intensive firms', *Journal of Management Studies*, **29** (6), 713–40.

Steedman, H. and K. Wagner (1989), 'Productivity, machinery and skills: clothing manufacture in Britain and Germany', *National Institute Economic Review*, (May), 41–57.

Stewart, R., J.L. Barsoux, A. Kieser, H.D. Ganter and P. Walgenbach (1994), *Managing in Britain and Germany*, New York: St. Martin's Press.

Strange, S. (1997), 'The future of global capitalism: or, will divergence persist forever?', in C. Crouch and W. Streeck (eds), *Political Economy of Modern Capitalism: Mapping Convergence and Diversity*, London: Sage, pp. 182–92.

Streeck, W. and R. Heinze (1999), 'An Arbeit fehlt es nicht', *Der Spiegel*, **19**, pp. 38–45.

Streeck, W. and K. Thelen (forthcoming), 'Institutional change in advanced political economies' in W. Streeck and K. Thelen (eds), *Continuity and Discontinuity in Institutional Analysis*, Oxford: Oxford University Press.

Sturdy, A. (1998), 'Strategic seduction? Information technology consultancy in UK financial services', in José Luis Alvarez (ed.), *The Diffusion and Consumption of Business Knowledge*, Basingstoke: Macmillan.

Swart, J. and N. Kinnie (2003), 'Knowledge intensive firms: the influence of the client on HR systems', *Human Resource Management Journal*, **13** (3), 37–55.

Tate, J. (2001), 'National varieties of standardization', in P. Hall and D. Soskice (eds), *Varieties of Capitalism: The Institutional Foundations of Comparative Advantage*, Oxford: Oxford University Press, pp. 442–73.

Teece, D.J. (1998), 'Capturing value from knowledge assets: the new economy, markets for know-how and intangible assets', *California Management Review*, **40** (3), 55–79.

Thelen, K. and C. van Wijnbergen (2003), 'The paradox of globalization: labor relations in Germany and beyond', *Comparative Political Studies*, **36** (8), 859–80.

Tomlinson, M. (2001), 'A new role for business services in economic growth', in D. Archibugi and B.A. Lundvall (eds), *The Globalizing Learning Economy*, Oxford: Oxford University Press, pp. 97–107.

Tordoir, P. (1995), *The Professional Knowledge Economy: The Management and Integration of Professional Services in Business Organizations*, Boston, MA: Kluwer Academic Publishers.

Tylecote, A. and E. Conesa (1999), 'Corporate governance, innovation systems and industrial performance', *Industry and Innovation*, **6** (1), 25–50.

Venkatraman, N. (1991), 'IT-induced business reconfiguration', in M. Scott Morton (ed.), *The Corporation of the 1990s: Information Technology and Organizational Transformation*, Oxford: Oxford University Press, pp. 122–58.

Vitols, S. (2001), 'Varieties of corporate governance: comparing Germany and the UK', in P. Hall and D. Soskice (eds), *Varieties of Capitalism: The Institutional Foundations of Comparative Advantage*, Oxford: Oxford University Press, pp. 337–60.

Weitzman, E. and M. Miles (1995), *Computer Programs for Qualitative Data Analysis: A Software Sourcebook*, Thousand Oaks, CA: Sage.

Whitley, R. (1999), *Divergent Capitalisms: The Social Structuring and Change of Business Systems*, Oxford: Oxford University Press.

Yin, R. (1981), 'The case-study crisis: some answers', *Administrative Science Quarterly*, **26** (1), 58–65.

Yin, R. (1994), *Case Study Research: Design and Methods*, 2nd edn, Newbury Park, CA: Sage.

Zentrum für Europäische Wirtschaftsforschung (ZEW) (1999), 'Germany's technological competitiveness', report prepared for the Federal Ministry for Education and Research, Mannheim.

Zimmermann, R. and S. Whittaker (eds) (2000), *Good Faith in European Contract Law*, Cambridge: Cambridge University Press.

7. Two types of organizational modularity: SAP, ERP product architecture and the German tipping point in the make/buy decision for IT services

Mark Lehrer

INTRODUCTION

The German IT sector of the 1990s witnessed exceptional changes in IT outsourcing patterns. Prior to this decade, IT outsourcing was comparatively underdeveloped among German companies (Lehrer 2000; Grimshaw and Miozzo Chapter 6). As a result, German firms were hardly represented among the top IT services companies in Europe (see Table 7.1).

Yet just a few years later the situation had changed substantially. Two major German IT service companies emerged (see Table 7.2). As Grimshaw and Miozzo (Chapter 6) document, Germany went seemingly overnight from a laggard in IT services (comparatively little outsourcing) to a leading country in IT services and outsourcing. Two major German IT service companies emerged (see Table 7.2). Why did the IT service industry develop so suddenly in Germany?

The focus here is on one piece of the puzzle, namely the role played by Systemanalyse und Programmentwicklung, or Systems Analysis and Program Development (SAP) and the massive installation of SAP's enterprise resource planning (ERP) systems by German firms. The simultaneity of SAP's success and increased German outsourcing in the 1990s is far from coincidental. ERP software package and external IT service provision were part of a common response to a new set of technological opportunities, notably new IT architectures (client/server networks) and the advent of hardware-independent operating systems (UNIX, Windows NT). Furthermore, companies that implemented ERP systems like SAP's R/3 software almost invariably relied on external IT consultants to implement the R/3 package. As explained below, ERP systems like R/3 require

Table 7.1   Largest IT service providers in West Europe, 1991

Rank	Firm	Market share	Country
1	IBM	20.2	USA
2	Cap Gemini Sogeti	6.1	F
3	BT Customer Systems	6.0	UK
4	DEC	4.5	USA
5	Andersen Consulting	4.3	USA
6	Sema Group	4.0	F
7	Logica	3.0	UK
8	Thomsen CSF (BSI)	2.6	F
9	Data Sciences	2.4	UK
10	**Siemens Nixdorf**	**2.1**	**D**
11	Olivetti	2.1	I
12	ACT Group	2.0	USA
–	Other	40.7	–

Source: Pierre Audoin Conseil, 1992.

Table 7.2   Largest IT service providers in West Europe, 1997

Rank	Firm	Turnover (€ million)	Country
1	IBM	5500	USA
2	EDS	3230	USA
3	Cap Gemini Sogeti	2530	F
4	Andersen Consulting	2040	USA
5	**Debis Systemhaus**	**1600**	**D**
6	Computer Sciences	1560	USA
7	**Siemens Nixdorf**	**1490**	**D**
8	Sema Group	1410	F
9	Bull	1280	F
10	Compaq/Digital	1050	USA
11	Finsiel	1010	ITA
12	Origin	970	NE
13	Oracle	950	USA
14	Atos	930	F
15	ICL	830	UK/JPN

Source: Pierre Audoin Conseil, 1998.

extensive customization by specialists familiar with the software package. In the 1990s, R/3 consultants were among the highest-paid consultants in the business.

This itself raises a paradox which this chapter will endeavour to address. Why did the leading developer of ERP systems, SAP, emerge in Germany? Why was the home market favourable to the global success of SAP? On the face of it, one would expect ERP systems to emerge later, not earlier in a country like Germany where in-house development of company IT systems had long been relatively preponderant. The more developed state of external IT markets outside Germany (as indicated in Table 7.1) would lead one to predict the leading vendor of ERP systems to emerge in countries like the USA, the UK or France where the IT service sector was more developed. Was there in the German IT sector a 'tipping point', a 'strategic inflection point' that caused German firms to switch en masse from predominately internal to increasingly external IT development? A look at the evolution and impact of SAP on the German IT sector helps provide answers.

ANALYTICAL TOOLS: TWO TYPES OF ORGANIZATIONAL MODULARITY

Central to the following analysis is the notion of organizational modularity. The rise and impact of SAP can be understood in terms of the modularity provided by ERP systems. This is not just a matter of modular product design. Rather, ERP systems affect the modularity of the organizations that implement them. We can distinguish two types of organizational modularity that will be relevant to our analysis of ERP systems: (1) market-organizational modularity and (2) function-organizational modularity.

Market-organizational modularity concerns the outsourcing phenomenon, specifically the decision by firms as to whether to purchase IT services on the external market or conduct these IT activities in-house. This is the make/buy decision as applied to business functions. As shown by Miozzo and Grimshaw (Chapter 4, this volume), market-organizational modularity is not an all-or-nothing affair: even when outsourcing their IT operations, firms generally retain an appreciable level of internal IT capabilities. Intersecting with theories of transaction costs (Williamson 1975, 1985) and vertical disintegration (Stigler 1951), market-organizational modularity refers to the way in which economic activity is partitioned among firms and markets. Sturgeon (2002) and Langlois (2002) are among recent authors who conceptualize the vertical disintegration of companies in terms of organizational modularity. The primary economic benefit of market-organizational modularity is generally considered to be Smithian specialization (Stigler 1951), although economies in information processing (Baldwin and Clark 1997) and credible commitments to markets (Chen 2002) are other cited benefits.

Function-organizational Modularity

This, in contrast, relates to the specific characteristics of ERP systems. An ERP system is essentially a company-wide core IT system that integrates the firm's different business functions. ERP systems are modular in the sense that they define interfaces among different function-specific components of the firm's IT system, thereby allowing IT changes in one business department to occur without disrupting the functionality of IT in other departments. Function-organizational modularity accommodates the dual need to orchestrate information-sharing among different departments and yet to allow sufficient independence of IT components so that local changes in one department do not require an overhaul of the entire IT system. In fact, the various components of SAP's flagship R/3 product are called 'modules.' A list of R/3 modules is included in Table 7.3.

The following analysis traces the relationship between these two types of organizational modularity in ERP systems. It will be seen, first, how the function-organizational modularity of ERP systems induced or at least accelerated a trend towards greater IT outsourcing, that is, towards greater market-organizational modularity. This sets the stage, second, for an analysis of why advanced ERP systems first emerged in Germany rather than in countries where IT outsourcing was more common.

ORGANIZATIONAL MODULARITY: THEORETICAL BACKGROUND

Market-organizational Modularity

The frequent isomorphism between product architecture and organizational structure has been noted in many contributions (Sanchez and Mahoney 1996; Brusoni and Prencipe 2001). At one level this is not surprising. Products are designed not only with a view to their functionality, after all, but also to the economy of production within firm processes and structures. Beyond this, however, once product designs are established, they tend to perpetuate existing organizational structures in production. In this case, 'architectural innovations' (Henderson and Clark 1990) may cause incumbent firms to fail when their organizational structures cannot adapt to changes in product architecture.

Table 7.3   SAP modules

(FI) Financial Accounting	(AM) Asset Management
(CO) Controlling	(PP) Production Planning
(PS) Project Systems	(QM) Quality Management
(FM) Funds Management	(WM) Warehouse Management
(MM) Materials Management	(OC) Office & Communications
(SD) Sales & Distribution	(HR) Human Resources

Sanchez and Mahoney (1996) claim that modularity in product design requires modularity in organization. In one sense, this kind of isomorphism is extremely common. Many manufactured products are assembled from components by final manufacturers who obtain the components from specialized suppliers. The set of suppliers that contribute to the final product constitutes a 'value network' (Christensen 1997). In industries as diverse as automobiles and personal computers, product modularity facilitates the outsourcing of component production; this can ultimately culminate in 'screwdriver' final assemblers who buy virtually all vital parts from outside suppliers (as in personal computers). The basic notion of value networks is essentially very old, traceable as both an empirical phenomenon and theoretical construct to the time of Adam Smith. In this spirit, Langlois (2002) associates organizational modularity largely with the market organization of industries, that is the coordination of industrial activity by market relations among many smaller firms rather than within large integrated firms.

Yet some authors claim that there is something new under the sun in recent modular designs and their corresponding supplier networks. Baldwin and Clark (1997) maintain that more sophisticated interface specifications in many industries enable assemblers to outsource not only the production, but also the very design of major components. The value of doing this is complexity reduction. By partitioning production tasks into visible design rules and hidden information (that is, details of component characteristics that need not concern other firms in the network as long as the visible design rules are adhered to), assemblers can delegate the design and improvement of components to more specialized suppliers without endangering overall product integrity.

Within ERP systems, such delegated design is crucial to the customization of the business software to the individual requirements of different corporate customers. ERP software is not a ready-to-use package, but rather an architecture with thousands upon thousands of parameters that need to be set by the user companies. ERP packages are comparable to operating systems, with much of the actual programming left to the user. In fact, one of the most useful tools supplied by SAP is its ABAP programming language with which user companies can write a software code to run tasks that are beyond the scope of the software's generic functionalities.

Function-organizational Modularity

Modularity consists of two faces. One face concerns the hierarchical decomposition of the overall product or task into units of lesser complexity that can be independently or semi-independently managed without any need for knowledge of the whole. Complex systems, biological or manmade, are generally organized in this way (Simon 1962). The other face of modularity concerns the integration of the components. As long established by Herbert Spencer, the Victorian biologist and early social philosopher, the greater the functional differentiation of constituent parts, the greater the need for functional integration.

The great value of ERP systems lies in the integration across functions they provide. By the 1980s, most companies had accumulated a patchwork of function-specific IT systems. That is, different company departments like finance, production, human resources, and accounting each possessed their own separate software systems (and often separate hardware as well). Prior to the advent of ERP, functionally differentiated IT systems within companies enjoyed one advantage of modularity – the insulation of the whole system from failure within a single subsystem – but failed to fulfil the integrative criterion of modularity. ERP vendors like SAP and PeopleSoft promised to remedy this with comprehensive IT systems that share and combine information across functions (which could be and usually were 're-engineered' in the process). The payoff to user firms is what Nightingale et al (2003) call 'economies of system': economies not deriving from scale or scope, but from the systemic linking of firm activities through large technical systems.[1] ERP systems were also an easy sell to top managers who were promised real-time access to all relevant company data and a means to overcome departmental information monopolies.

Codified knowledge is central to technical interfaces. In modular products, interfaces are explicit and often highly detailed. ERP implementation requires extensive codification of interfaces among business functions. In fact, the planning of an ERP system generally requires the firm to engage in an often unprecedented level of business process codification (Wahl 2003). This is a major reason for the nexus between ERP implementation and IT outsourcing; external consultants are needed not just for parameter-setting and supplementary software coding, but for business modelling and mapping out of desired business practices.

The diffusion of ERP systems in the 1990s thus entailed the participation of consulting companies in codifying firm processes (Schwarz 2000). Whatever the merits of tacit knowledge may be, planning and implementing an ERP system indubitably involves the transformation of semi-tacit into highly codified knowledge; only a detailed and formal modelling of business practices can identify all the relevant company information that needs to be processed and shared across different organizational units (Wahl 2003). This is accompanied by homogenization of technical terms across the company and the formulation of harmonized user interfaces within the firm's IT system.

GENERALIZED COMPETENCE AND THE DEVELOPMENT OF EXTERNAL MARKET CAPABILITIES

The Link between Market-organizational and Funtion-organizational Modularity

The growth of ERP systems and IT outsourcing highlights the evolution of a capability from one primarily organized within firms to one increasingly organized by external markets. We have already seen why a functionally integrative IT product like ERP systems led to greater involvement of IT service firms in ERP implementation. We now seek to explain why and how the function-organizational modularity of ERP systems was first developed by an independent software vendor (SAP; later followed by Baan and PeopleSoft) rather than by user firms themselves.

As a first step in theoretical generality, it is worth noting that ERP systems, like company IT infrastructure in general, are capital goods. The mid-nineteenth-century development of capital goods, specifically machine tools as traced by Rosenberg (1972, 1976), reveals curious parallels to the evolution of company IT systems in the 1990s. Before 1850, most US manufacturers had to produce their own machine tools in addition to their final products. But at just the time when the 'American system of manufactures' with its innovation of interchangeable parts was attracting international attention at the Crystal Palace exhibition of 1851, US industry had begun a course of vertical disintegration. Machine tools became an increasingly distinct market sector from which final manufacturers purchased their capital equipment. Interestingly, such vertical disintegration developed less extensively in the UK where final manufacturers remained wedded to highly customized specification for machine tools (Rosenberg 1972).[2]

The machine tools example shows how market-organizational modularity is fostered by standardization. According to Rosenberg, US consumers accepted guns, cutlery, cloths and stoves that were more functional and standardized than European counterpart goods. This enabled the technology of interchangeable parts to diffuse and 'lower the cost of innovation throughout the metal-using sectors of the economy' (Rosenberg 1972, p. 102). For example, automatic turret lathes originated in firearms production but spread to sewing machines (1850s), bicycles (1890s) and automobiles (1900s), to name but a few industries. The key result of this 'technological convergence' (Rosenberg 1976, p. 16) across industries was the emergence of 'general-purpose machinery' in place of specialized, industry-specific machinery. Rosenberg and Trajtenberg (2001) extend this to the concept of a 'general purpose technology' and analyse the parallel between the nineteenth-century Corliss steam engine (with its universally useful self-regulating rotary motion) and the semiconductor (with its universally useful binary logic system).

The rise of ERP systems can be similarly described as the evolution from a specific to a general-purpose capital good, in this case business software. Until the 1990s, business processes were assumed to be too idiosyncratic and variable from one company to another to permit large-scale software standardization except for very specific tasks. Standard software packages for business were prevalent after 1969 (the date of IBM's court-mandated unbundling of hardware and software), but these programmes were function-specific (accounting, payroll and so on). Indeed, all major ERP vendors originally started out selling function-specific software: SAP specialized originally in accounting and purchasing, PeopleSoft in human resources, and Baan in manufacturing (Meissner 1997, pp. 86–7; Rashid et al. 2002).

When SAP's R/3 product appeared as a general-purpose technology that could span and integrate all the functions of medium-to-large companies, there was nothing like it on the market. The next section describes the development of SAP's hit product and explores the evolution of business software from a specific-purpose to a general-purpose technology.

THE DEVELOPMENT OF STANDARD BUSINESS SOFTWARE AT SAP, 1972–99

SAP was founded in Walldorf, Germany in 1972 by four ex-IBM programmers. At the time, computers were so expensive that SAP had to conduct all of its product development on customer premises; SAP did not even own a computer until 1980 (Plattner 2000, p. 26). From the beginning,

SAP's objective was to economize on programming costs by developing a standardized software code that could be partially reused as contracts migrated from one customer to another. This strategy had to be partially concealed, however, for the market at the time demanded that 'you have to offer users something specifically tailored to them' (firm founder Dietmar Hopp, cited in Meissner 1997, p. 48). Each new contract was used to enlarge the scope of the standardized core code.

As a result, SAP's business software acquired an increasing level of generality over time. This generality necessarily entailed an increasing level of complexity and abstraction as well, with SAP software incorporating an expanding range of adjustable parameters to fulfil specific customer requirements. Originally written for IBM computers, SAP's code was ported in the early 1980s onto a common platform so that it could be installed on the mainframes of Siemens and other manufacturers (Meissner 1997, pp. 50–1).

The first major product suite of multiple business functions was R/2, completed in 1981. It was designed for mainframe systems using timeshare terminals. In the 1980s, R/2 quietly conquered the German market for large-business software on mainframes. The architecture of R/2 was attractive in two major aspects. First, it was constructed on a centralized data bank that obviated the need to enter records more than once. Data could be entered on decentralized computer terminals via standardized presentation menus (the 'R' stands for 'real-time' in contrast to batch processing). Second, the various functional modules were both separable and interoperable. This enabled firms to implement the various modules (finance, accounting, production, and so on) sequentially rather than all at once. Once installed, cross-functional integration among functions was facilitated by a common architecture and database. The IT magazine *Computerwoche* noted in 1990: 'Almost unnoticed by the general public, SAP AG of Walldorf has acquired a quasi-monopoly in commercial software for IBM-370 computers in the Federal Republic with the *modular* standard software package R/2' (Meissner 1997, p. 53, emphasis added). Already in the 1980s, SAP found it necessary to delegate implementation of its software to consulting and IT service companies. SAP's connections to IT service companies as well as to the Big Six accounting companies, vital to SAP's later worldwide expansion, date from the R/2 era (Plattner 2000, p. 38).

SAP's follow-up product, R/3, formed the basis of its ERP market leadership. It replaced the IBM-based architecture of R/2 with UNIX-centred client/server architecture. Client/server architecture meant that R/3 was scalable; the size of the company IT system could be augmented at will by adding servers and clients. When the power of PCs and PC networks exploded in the 1990s, R/3-based systems became more powerful than mainframe-based R/2 systems. At the dawn of the PC and Internet boom,

SAP found itself with a virtual monopoly on modular ERP systems that could run on a client/server architecture. Though originally conceived specifically for medium-sized firms, R/3 was eagerly sought by global companies looking to integrate their cross-border operations on increasingly standardized server and PC hardware.

When SAP introduced R/3 in 1991, no-one could foresee the potential of client/server networks. The global take-off of R/3 was as serendipitous as its development. SAP had originally conceived R/3 in 1987 to comply with the new architectural concept of IBM called System Application Architecture (SAA). Development of R/3 for SAA brought SAP into the realm of UNIX, relational databases, and the programming language C, all easily adaptable to the later hegemony of PC-based client/server networks. Of particular importance, UNIX was a non-proprietary operating system that could run on a multitude of hardware platforms. When the IBM hardware on which R/3 was to make its debut failed to meet performance expectations, SAP scrambled to port R/3 onto a puny UNIX workstation (Plattner 2000, p. 33). Given the contrast in size to the huge mainframe systems that had formerly hosted SAP software, the workstation-based installation of R/3 caused a sensation at the 1991 CeBIT fair. By 1992 SAP was able to roll out R/3 on UNIX machines independent of any manufacturer; the conversion to Windows NT in 1993 was a simple affair, requiring only five days of coding (Plattner 2000). R/3 took off: in the years 1990–99, SAP's turnover increased 20-fold from DM500 million to over DM10 billion (€5 billion).

The evolution of SAP's standard business software from a specific-purpose to a general-purpose technology therefore consisted of two strands. First, the basic code obtained an ever higher level of functional generality and abstraction with each installation; the wider the range of businesses and industries that SAP had to serve, the greater the number of adjustable parameters SAP's business software included. Second, SAP increasingly liberated its programming code from IBM mainframe standards; in the 1990s, hardware specifications ceased to be an important constraint.

The development is comparable with that of the US firm of Brown and Sharpe in the 1800s. Originally a producer of clocks and watches, the firm's foray into sewing machines induced it to develop a novel turret screw machine. This machine was quickly purchased by firms across many industries, including tools, shoe machinery, locomotives, rifles, and machine tools themselves. Similar success was encountered in Brown and Sharpe's precision grinding and universal polishing machines; the latter were originally developed for weapons makers, but quickly found a vast market covering over a dozen industries, including machine tools themselves (Rosenberg 1976, p. 23). As in the case of SAP, Brown and Sharpe engaged in the refinement of capital goods equipment originally designed to solve industry-specific problems but which quickly found application across a range of industries.

Technical Complexity: the Link between Function-organizational and Market-organizational Modularity

Clearly, R/3 is hardly plug-and-play software, requiring instead extensive customization to each customer's premises. In practice, R/3 implementers have to parameterize thousands of tables over a period of 2–4 years and write time-consuming supplemental software codes. The cost of implementing the software generally runs above five times the purchase price of the basic software licence from SAP (Cooke and Peterson 1998; Schwarz 2000, p. 27). The complexity of the undertaking exceeds the capability of company IT departments, thus requiring SAP or (more often) an external IT service or consulting company to supervise implementation of the ERP system. It is the technical complexity surrounding ERP systems, then, that constitutes the primary causal link between function-organizational modularity (the nature of the R/3 software product) and market-organizational modularity (the delegation of ERP system development to specialized IT service firms). The complexity of ERP systems entailed market organization of both the basic software product (SAP, PeopleSoft, Baan, and so on) *and* of its implementation (by IT service providers).

The market for ERP systems grew rapidly in the 1990s. In Western Europe, estimated outlays for ERP systems doubled from 1995 to 1998 alone, from 1.5 to 3 billion euros; spending on ERP grew at three times the 10 per cent per annum increase in overall IT turnover in these years.[3] Aided by the penetration of R/3 (which went hand in hand with the corporate transition to client/server networks) IT outsourcing in Germany made rapid strides in the 1990s.[4] At just the time when most of Germany's medium to large computer companies went bust (Siemens-Nixdorf exited hardware, then software as well), larger German industrial conglomerates discovered they were able to compete in IT service provision. With SAP installations as a major driver of business, the IT service divisions of Debis (of Daimler-Chrysler) and Siemens grew dramatically in the 1990s, to the point of joining the ranks of the largest European IT service firms (see Table 7.2 above).

German IT service providers also began to manifest an increasing international presence. SAP-specialized service providers like CSC Ploenzke and Plaut took advantage of the SAP boom to expand into foreign countries. The SAP services divisions of Debis and Siemens-Nixdorf likewise expanded rapidly abroad. Makers of complementary software products, like Ixos Software AG and IDS Prof. Scheer, also grew rapidly in the 1990s, riding on the coat-tails of SAP.

THE GERMAN IT OUTSOURCING PARADOX: A MACRO INSTITUTIONAL EXPLANATION

To return to the question raised at the beginning, why did Germany go from being a laggard to a leader in IT outsourcing? German companies jumped seemingly overnight onto a bandwagon of 'buying' rather than 'making' IT services. While the activities of SAP are an important intermediate link in the causation chain, it remains to be explained why German companies initially relied relatively more on in-house software development than elsewhere, why the leading ERP provider emerged in Germany rather than elsewhere, and why there was a sudden rather than gradual movement from in-house to outsourced IT development in Germany. In more theoretical terms, why was the German business environment propitious to function-organizational modularity in business software (ERP systems) and subject to a 'tipping point' in reliance on IT outsourcing (market-organizational modularity)?

The mass penetration of SAP software into large German companies during the 1980s reveals that reliance on external IT expertise is nothing new in Germany. If nonetheless the IT outsourcing market prior to the 1990s was underdeveloped and if Germany lacked the large internationally operating IT service companies found in the USA, UK and France (see Table 7.1, above), the inference to be drawn is that German companies preferred to maintain strong control over routine IT operations while relying on external IT expertise for development of specific IT infrastructure. The preference for in-house control over IT made sense within the institutional context of German employment (Grimshaw and Miozzo, Chapter 6 this volume). The institutional basis of German business, from training and union organization to labour law and pension plans, favours long-term employment relations within firms (Streeck 1997; Hall and Soskice 2001). This is more than just a paternalistic tradition of lifetime or long-term employment. Both the hiring and termination of employees entail high direct and indirect costs for German employers. The celebrated commitment of German industry to investing in human capital therefore has a fairly straightforward economic rationale (Soskice 1994), which is not to deny the competitive and social merits of Germany's extensive training systems predicated on long-term employment.

Yet the 'employment effect' is only one effect of German business institutions on corporate IT decisions. There is another effect of these institutions that prepared the ground for the diffusion of ERP systems within Germany and somewhat attenuated the employment effect. This counterbalancing force is the 'standardization effect'. Beyond Germany's famous system of DIN norms, German industrial practice is highly coordinated by influential industry and labour organizations whose net effect is to homogenize many work and business practices. Such sector-based coordination, extending far beyond mere wage bargaining, involves widespread standardization of training, work processes and job classifications (Lane 1989; Hall and Soskice 2001). Cross-national comparisons of industries regularly reveal greater work-process homogeneity among German firms than among US or UK firms, for example (Berg 1994; Lane and Bachmann 1996). This standardization is reinforced by the interplay between German business institutions and Germany's industrial specialization. Post-war Germany has come to be increasingly specialized in medium-tech industries where change is incremental rather than radical (Casper et al. 1999; Siebert and Stolpe 2002) and where work process standardization is therefore feasible; this specialization can be largely explained by the institutional restrictions imposed on firms by the domestic political context.

The interfirm standardization of business processes favours the implementation of standardized software products that can be transferred from one firm to another. In essence, SAP benefited from a domestic market in which firm processes were more standardized and, in all likelihood, more explicitly codified than in other countries. While it may be tempting to attribute such standardization and codification to cultural stereotypes of Germanic methodological planning, evolutionary explanations are more plausible. Given the heavy restrictions of German socio-political institutions, it is probable that only the most methodically organized enterprises could prosper within the domestic environment. In other words, the method-driven and planning-obsessed companies so reported by foreign observers to Germany are most likely the survivors of environmental selection pressures that also happened to be of benefit to software developer SAP. Enterprise-wide ERP systems emerged first in Germany because it was easier to standardize business software within the German context.

The diffusion of R/2 and R/3 hastened the trend towards IT outsourcing because of the cost and technical complexity of implementation. Given such cost and complexity, implementation expertise could be more economically accumulated in outside IT firms where it could be easily redeployed. Through a virtuous cycle, once such implementation expertise was accumulated in external IT service providers and a certain regularity in the provision of such services prevailed, the German market for IT services 'thickened', lowering the uncertainty of relying on external vendors. In sum, SAP's standard software product helped thicken the market for IT service provision and thereby increased the relative advantages of 'buy' over 'make' in the development of business software.

An additional impetus to the German standardization effect came from technological factors of the 1990s that were not specific to Germany. Prior to the 1990s, most companies relied on a motley mix of hardware and software programs accumulated over the years. The advent of client-network systems and the threat of the Y2K bug provided firms with a heavy incentive to overhaul their IT systems completely and optimize ('re-engineer') their business processes at the same time. Rapid performance improvements and increasing standardization of PCs and client/server architectures favoured the development of new markets for standardized business software and associated external IT services. By liberating firms from proprietary hardware standards, client/server and PC networks allowed unified business software to run on a composite of hardware systems. ERP systems were less a cause than a catalyst for overhauling corporate IT systems and designing a more highly 'rationalized' company IT architecture.

The tipping point in German IT outsourcing is represented in Figure 7.1. Up to a certain level of technical complexity, the employment effect

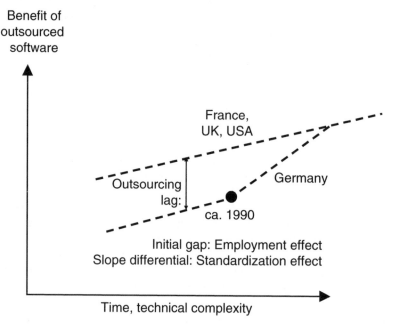

Figure 7.1   A model of the German IT tipping point

made in-house IT development comparatively more attractive to German companies than to French, US, or UK companies. After a certain level of technical complexity was reached, however, the standardization effect raised the relative advantages of purchasing packaged business software and then relying on external IT service providers for its implementation. Figure 7.1 is admittedly impressionistic. For many large German companies the tipping point began in the 1980s when R/2 was available, while for medium-sized as well as many multinational companies the tipping point came in the 1990s with the near-simultaneous arrival of R/3 and client/server architectures.

CONCLUSION

We are now in a position to return to the question raised at the beginning. Why did Germany go seemingly overnight from an undersized IT outsourcing market to a major IT outsourcing market? Why was there a 'tipping point' in the propensity to outsource IT? Why was there discontinuity as opposed to continuity?

The answer seems to lie mainly in the realm of technological supply. The PC revolution and the emergence of client/server architectures reshuffled IT markets. Increasingly freed from the technical specifications of mainframe manufacturers, software and hardware could be mixed and matched in entirely new ways. Client/server networks, with their famous scalability, allowed firms to reconfigure their corporate IT infrastructure as modular systems, with separate departmental capabilities linked through common interfaces, standardized terminology and a shared repository of information (usually a relational database).

The new technology supply wave of the 1990s offered many possibilities but also presented firms with a far higher level of complexity to manage; firms found it increasingly unfeasible to depend on large computer companies or on their own IT departments to figure out the optimal mix of hardware and software systems that suited them. This dilemma was actually deepened, not lessened by the IT supply shock that was felt strongest in Germany: the availability of a comprehensive business software package, SAP's R/3, that could run on client/server networks. For although R/3 offered the potential of substantial performance increases over existing business software, the costs and complexity of implementation were extremely high.

The supply idiosyncrasy of the German IT sector lay in business software and, as we have seen, it began prior to the 1990s movement towards client/server architectures. SAP began installing comprehensive business software packages in the 1970s and 1980s, and according to SAP's CEO Hasso Plattner, only SAP and other *German* competitors offered business software with cross-functional integration prior to the 1990s (Plattner 2000, p. 39). This supply idiosyncrasy, we have suggested, arose in conjunction with demand idiosyncrasies of German user firms, in particular the greater interfirm standardization of business processes. This can be ascribed to the macro institutional factors discussed above.

Even in the absence of SAP, there would have been a tipping point in German IT outsourcing. After relying more on in-house IT development than in other countries, German firms in the 1990s would have been compelled in any case to 'catch up' with other countries in the propensity to outsource IT. The global rise of SAP accompanied rather than caused the greater trend towards IT outsourcing because both phenomena were responses to a common set of technological changes and opportunities. Without a doubt, however, the presence of the world's leading ERP provider also had two additional ripple effects on the German IT sector. Because SAP's forerunner R/2 was already so well known in Germany, the successor R/3 diffused quickly there, accelerating demand for supplementary external IT expertise. Second, German IT service and consulting firms that had worked with SAP software suddenly found themselves with a comparative knowledge advantage that they could leverage by expanding abroad.

NOTES

1. Although cross-functional synergies were the original selling point of ERP systems, they subsequently enabled synergies across different businesses and national subsidiaries as well, as mentioned below.
2. The US–UK comparison is important because it shows that market size alone – with its expected Smithian effects on specialization – could not have been the only explanatory variable. Because the UK economy remained larger than the US economy for most of the nineteenth century, the related hypothesized correlation between vertical disintegration and market size (Stigler 1951) fails to provide a completely adequate explanation.
3. Source: International Data Corporation (IDC), several issues.
4. The vast majority of SAP's German customers who had implemented R/2 in the 1980s migrated to R/3 in the 1990s, usually availing themselves of the opportunity to engage in re-engineering, incorporate Internet technology and/or fix the Y2K bug.

REFERENCES

Baldwin, C. and K. Clark (1997), 'Managing in an age of modularity', *Harvard Business Review*, **75** (5), 84–93.

Berg, P. (1994), 'Strategic adjustments in training: a comparative analysis of the US and German automobile industries', in Lisa Lynch (ed.), *Training and the Private Sector: International Comparisons*, Chicago: University of Chicago Press, pp. 77–107.

Brusoni, S. and A. Prencipe (2001), 'Unpacking the black box of modularity: technologies, products and organization', *Industrial and Corporate Change*, **10** (1), 179–205.

Casper, S., M. Lehrer and D. Soskice (1999), 'Can high-technology industries prosper in Germany? Institutional frameworks and the evolution of the German software and biotechnology industries', *Industry and Innovation*, **6** (1), 5–24.

Chen, Y. (2002), 'Vertical disintegration', working paper, University of Colorado at Boulder.

Christensen, C. (1997), *The Innovator's Dilemma: When New Technologies Cause Great Firms to Fail*, Cambridge, MA: Harvard Business School Press.

Cooke, D. and W. Peterson (1998), *SAP Implementation: Strategies and Results*, New York: Conference Board.

Hall, P. and D. Soskice (eds) (2001), *Varieties of Capitalism: The Institutional Foundations of Comparative Advantage*, Oxford: Oxford University Press.

Henderson, R. and K. Clark (1990), 'Architectural innovation: the reconfiguration of existing product technologies and the failure of established firms', *Administrative Science Quarterly*, **35** (1), 9–30.

Lane, C. (1989), *Management and Labour in Europe: The Industrial Enterprise in Germany, Britain and France*, Aldershot, UK and Brookfield, US: Edward Elgar.

Lane, C. and R. Bachmann (1996), 'The social constitution of trust: supplier relations in Britain and Germany', *Organization Studies*, **17** (3), 365–95.

Langlois, R. (2002), 'Modularity in technology and organization', *Journal of Economic Behaviour and Organization*, **49** (1) 19–37.

Lehrer, M. (2002), 'Has Germany finally fixed its high-tech problem? The recent boom in German technology-based entrepreneurship', *California Management Review*, **42** (4), 89–107.

Meissner, G. (1997), *SAP – Die heimliche Software-Macht*, Hamburg: Hoffmann und Campe.

Nightingale, P., T. Brady, A. Davies and J. Hall (2003), 'Capacity utilization revisited: software, control and the growth of large technical systems', *Industrial and Corporate Change*, **12** (3), 477–517.

Pierre Audoin Conseil (1992), *Survey: Software and IT Services in Europe*, Paris: Pierre Audoin Conseil.

Pierre Audoin Conseil (1998), *Survey: Software and IT Services in Europe*, Paris: Pierre Audoin Conseil.

Plattner, H. (2000), *Dem Wandel voraus*, Bonn: Galileo Press.

Rashid, M., L. Hossain and J. Patrick (2002), 'The evolution of ERP systems: a historical perspective', in L. Hossain, J. Patrick and M. Rashid (eds), *Enterprise Resource Planning: Global Opportunities and Challenges*, Hershey, PA: Idea Group Publishing, pp. 1–16.

Rosenberg, N. (1972), *Technology and American Economic Growth*, New York: Harper.

Rosenberg, N. (1976), *Perspectives on Technology*, Cambridge: Cambridge University Press.

Rosenberg, N. and M. Trajtenberg (2001), 'A general purpose technology at work: the Corliss steam engine in the late 19th century US', working paper, CEPR.

Sanchez, R. and J. Mahoney (1996), 'Modularity, flexibility, and knowledge management in product and organization design', *Strategic Management Journal*, **17** (special Winter issue), 63–76.

Schwarz, M. (2000), *ERP-Standardsoftware und organisatorischer Wandel*, Wiesbaden: Deutscher Universitäts-Verlag.

Siebert, H. and M. Stolpe (2002), 'Germany', in B. Steil, D. Victor and R. Nelson (eds), *Technological Innovation and Economic Performance*, Princeton, NJ: Princeton University Press, pp. 112–47.

Simon, H. (1962), 'The architecture of complexity', *Proceedings of the American Philosophical Society*, **106**, 467–82.

Soskice, D. (1994), 'Reconciling markets and institutions: the German apprenticeship system', in L. Lynch (ed.), *Training and the Private Sector: International Comparisons*, Chicago: University of Chicago Press, pp. 25–60.

Stigler, G. (1951), 'The division of labour is limited by the extent of the market', *Journal of Political Economy*, **54** (3), 185–93.

Streeck, W. (1997), 'The German economic model: does it exist? Can it survive?', in C. Crouch and W. Streeck (eds), *Political Economy of Modern Capitalism: Mapping Convergence and Diversity*, London: Sage, pp. 33–54.

Sturgeon, T. (2002), 'Modular production networks: a new American model of industrial organization', *Industrial and Corporate Change*, **11** (3), 451–96.

Wahl, M. (2003), *Wissensmanagement im Lebenszyklus von ERP-Systemen*, Wiesbaden: Deutscher Universitäts-Verlag.

Williamson, O. (1975), *Markets and Hierarchies: Analysis and Antitrust Implications*, New York: Free Press.

Williamson, O. (1985), *The Economic Institutions of Capitalism*, New York: Free Press.

8. Managing competencies within entrepreneurial technologies: a comparative institutional analysis of software firms in Germany and the UK

Steven Casper and Sigurt Vitols

INTRODUCTION

In recent years the US economy has developed an institutional environment fostering the widespread use of entrepreneurial business models to support clusters of dynamic small firms specializing in new technologies. How do differences in national business system frameworks influence the development of new technology firms? Do business strategies and related organizational structures developed by European firms simply mimic those found in the USA, or have European firms found unique organizational formulas for translating technology investments into commercial enterprises? From a theoretical perspective, is comparative institutional theory helpful in examining the adaptation of new organizational forms across different economies?

This chapter develops and empirically tests a theoretical framework to evaluate the impact of national institutional frameworks on the organization and innovation strategy of entrepreneurial technology firms in the software industry. It then develops and tests a number of empirical hypotheses linking the orientation of national institutional frameworks to the successful governance of organizational dilemmas facing different types of entrepreneurial firms in the UK and Germany. Using a cluster analysis, the chapter empirically demonstrates the existence of distinct types of entrepreneurial firms within the software industry, and then shows that national patterns of specialization across these firm types are influenced by the orientation of national institutional frameworks.

We draw on recent theoretical literature in the field of 'varieties of capitalism' (Casper et al. 1999; Hall and Soskice 2001; Hollingsworth

1997; Whitley 1999) to motivate our analysis. A core assertion of the varieties of capitalism approach is that national patterns of specialization are created by comparative institutional advantages in managing the organizational competencies needed to innovate within particular technological fields. This framework predicts that national institutional structures associated with 'liberal market economies' (LMEs) (e.g. the USA, the UK, or Canada), such as deregulated labour markets, capital-market-based financial systems, and shareholder-primacy-oriented company law, support the flexible orchestration of competencies needed to perform well in quickly changing or 'radically innovative' industries. However, these structures do not advantage the governance of longer-term, process-oriented innovation strategies in which longer-term human resource organization and 'patient' finance are needed. Due to their more regulated labour markets, organized training system, and bank-centred financial system, Germany and Sweden, on the other hand, are seen as 'coordinated market economies' (CMEs) that can promote performance in industries characterized by incremental or process innovations.

We generate several hypotheses linking national institutional characteristics in Germany and the UK to patterns of industry specialization in the software industry. To test these hypotheses, a cluster analysis is performed to analyse data on 190 software firms listed on stock exchanges in Germany and the UK. If firms cluster into stable groupings that represent distinct sub-sectors within the software industry *and* if the distribution of the nationality of firms across clusters is correlated with patterns of sub-sector specialization associated with particular national institutional frameworks, then this will help confirm the theory that national institutional factors influence patterns of sub-sectoral industry specialization within new technology industries such as software. Our evidence supports both conjectures.

The chapter is organized into three sections followed by a conclusion. The first section develops a theoretical model exploring differences in the technological, market, and organizational risks facing different types of entrepreneurial technology firms, with reference to the software industry. Focusing on Germany and the UK, the second section examines how the orientation of institutional frameworks associated with different 'models of capitalism' influences the governance of different types of technology firms, and develops several hypotheses. The third section describes the cluster analysis developed to test these hypotheses. This is followed by a conclusion summarizing results and highlighting the implications of this study for both institutional theory and the debate concerning the diffusion of entrepreneurial technology models within Europe.

MANAGING RISKS WITHIN ENTREPRENEURIAL TECHNOLOGY FIRMS

How do institutional frameworks impact patterns of competency formation within the software industry? We draw on concepts from research on 'sectoral systems of innovation' (Malerba and Orsenigo 1993; Mowery and Nelson 1999) to illustrate how firms within the software industry face different constellations of technological and market risks. While much research has generalized around entire sectors, such as biotechnology, machine tools, or software, we focus on important differences across different *sub-sectors* of the software industry. Later, when introducing institutional arguments, this leads to predictions linking the orientation of institutional frameworks to advantages in governing managerial risks associated with particular market segments. We focus on three types of risk: technological, market, and organizational.

'Technological risk' concerns the ability of a firm to develop capabilities needed to successfully pursue its chosen research and development (R&D) path (Perrow 1985; Woodward 1965). Problems stemming from the rate of technological change, or cumulativeness, form the main technological risk for most entrepreneurial technology firms (see Breschi and Malerba 1997). Viewed in terms of company capabilities, cumulativeness relates to the rate by which specific technological assets change during the evolution of an industry. If cumulativeness is low, this implies that Schumpeterian patterns of 'competency destruction' are high within an industry. Particular technological competencies within a firm have a high probability of failing (as they are shown to be inappropriate for resolving particular R&D problems). Firms in industries where cumulativeness is low often fail or, if they have sufficient financial resources to do so, must develop a capacity to quickly adjust their technological assets.

'Market risk' is defined as the ease by which firms can capture value from innovations, or appropriability. Following Teece (1986), appropriability regimes may be regarded as 'tight' when a firm is able to protect an innovation from being mimicked by competitors – typically through either standard forms of intellectual property protection or through trade secrets – thereby earning rents directly from the assets used to innovate. When appropriability regimes are weak, technological assets developed by the firm are difficult to safeguard and may be easily mimicked by competitors. In this case Teece has suggested that, to capture value from innovation, firms must develop complementary assets that are both specific to the firm and can be co-specialized or tied to generic assets. Viewed in terms of company organization, we will argue that firms developing co-specialized assets tend to create more complex organizational structures than firms innovating within tight appropriability regimes.

'Organizational risk' is defined as uncertainty surrounding the ability of a firm to effectively recruit, organize, and create incentives needed for personnel to innovate given particular types of technological and market risks facing the firm. While technological and market risks are exogenous to a firm, the ability to develop adequate organizational competencies is directly under managerial control. In other words, we define particular innovation systems through their identification with persistent patterns of technological and market risk that jointly creates a constellation of organizational challenges facing firms. Competitive success is associated with the ability of a firm to develop effective organizational competencies to manage these risks at a lower governance cost than competitors (Milgrom and Roberts 1993).

Through matching different combinations of technological and market risk, Figure 8.1 creates a typology of different types of entrepreneurial technology firms. We now examine each cluster of technological and market risk in more detail, illustrating how relatively distinct organizational solutions to these problems – i.e. types of firm – exist. We also suggest that well-known segments of the software industry correspond to each ideal-typical type of entrepreneurial firm.

Partnerships

'Partnerships' are the simplest form of entrepreneurial technology firm. Cumulativeness is relatively high, while most R&D can be readily appropriated by the firm. Relatively high technological cumulativeness entails that

		Technological Risk	
		Low	High
Market	Low	No managerial risks *Partnerships*	Competency destruction risks *Project-based firms*
Risk	High	Managerial hold-up risks *Collaborative firms*	Managerial hold-up risks Competency destruction risks *Basic research organizations*

Figure 8.1   A typology of entrepreneurial technology firms

failure risks are low, implying that knowledge investments made by employees are not risky (particularly when technical skills are common across a sector). Entrepreneurial firms with these characteristics generally resemble partnerships of highly skilled engineers or technicians that perform extensive technological consultancy, implementation, or customization work for particular clients. The partnership model, as developed within management consultancies, law firms, or medical practices, is common for this type of entrepreneurial firms. Because the need for managerial coordination is relatively low, ownership gravitates to the holders of technical expertise within the organization. Firms emerge primarily to develop brand awareness or develop platforms of specific competencies that can be bundled together for particular clients.

Within the software industry, information technology (IT) service firms fall within this category. IT service firms generally do very little in-house R&D, but instead hire teams of consultants and technical experts that do extensive customization work for clients using standardized tools and third-party software. Systems integration work is the most long-established market segment for IT service firms. Another involves enterprise software installations (ERP, CRM, or human resource systems) for clients for which enterprise software vendors (i.e. SAP, Oracle, Peoplesoft) prefer not to perform in-house customization work. IT service firms in this category often develop sector-specific expertise (i.e. within the finance sector) or service small firms for which the ERP vendors do not wish to provide in-house service. A final, rapidly growing segment of IT services is Internet-related consulting. IT service firms in this category usually provide a combination of e-commerce strategy consulting and web-development work. Because web-authoring tools and languages have become standardized, work again consists almost entirely of customization activities for clients performed by teams of consultants, designers, and web programmers.

Project-based Firms

'Project-based firms' resemble 'radically innovative' start-ups commonly associated with Silicon Valley and other technology clusters (see Saxenian 1994). Project-based firms specialize in sectors for which appropriability regimes for successful innovations are strong and in which innovations can be marketed without extensive customization for clients. This simplifies the organizational structure of firms, facilitating a focus on R&D with little need to develop and integrate marketing and distribution assets with core R&D. Through selecting segments in which market risks are low, project-based firms can focus on innovative activities with high technological risks created by low cumulativeness. Firms typically race against one another to develop new technologies that are capable of capturing large markets. Rents are generated either through licensing strategies based around patent protection or by developing products that could eventually become dominant designs within an industry (Utterback 1996), creating a combination of network externalities or customer lock-ins that can produce long-term rents for successful innovators. In either case, firms are organized to produce 'radical innovations' before competitors can capture large markets. However, many markets populated by project-based firms have a 'winner take all' character; low entry barriers create high risks that the firm will fail to innovate before its competitors.

Managers of project-based firms must develop powerful incentives to encourage highly skilled scientists and engineers to commit to what are often extremely challenging workplace environments. A core problem is the management of career risks created by a high probability of competency destruction. Because high technological risk denotes a likelihood of either outright failure or rapidly changing R&D trajectories entailing 'hire and fire' personnel policies, skilled employees may refuse to work within a firm if doing so poses a high risk of unemployment or a risk that a large percentage of skills acquired while working within the firm are not saleable on open labour markets. Project-based firms are usually the product of venture capital finance, and as such use a variety of equity-based incentive structures to develop and align performance incentives between investors, managers, and key scientists or engineers. A second strategy to minimize this problem is for the management of project-oriented firms to work with industry-wide rather than firm-specific technical skills whenever possible. This is a common strategy employed by new technology firms in many areas of biotechnology and, as discussed below, parts of the software industry. The development of industry-specific skills facilitates the development of networks of scientists and engineers working within particular technology niches, often located within regional clusters. Participation within such technological communities can dramatically lower the career risk of particular employees working on a given project while facilitating inter-firm collaboration (Powell 1996) within the software industry, standard (or application-based) software firms share characteristics of project-based organizations. Standard software is created for homogenous markets where the need for customization is low. Examples include graphic application software (e.g. CAD/CAM), multimedia and computer entertainment software, and a variety of application software used to run computer networks (e.g. e-mail, FTP, groupware, and document management programs). Intellectual property for software has traditionally been relatively weak. While copyright laws can protect a program's source-code, the 'look and feel' of a product can be mimicked by competitors (see Mowery 1999). Nevertheless, the nature of product market competition and the extremely low marginal cost of manufacturing successful standard software products can create large profits for successful innovators. Within many business or network-impacted markets a combination of network externalities and end-user lock-in effects creates large markets for successful innovators (see Shapiro and Varian 1999). Switching costs are lower in other segments of the standard software market, such as computer games or multimedia software. Large consumer markets for these products (leading computer games routinely sell several million copies) ensure high profitability for successful firms. Relatively low entry barriers within consumer markets and extremely high long-term profitability for successful firms in network or business-impacted segments leads to intensive innovation races across standard software firms. This generates high technological volatility as rivals race to introduce new features or 'functionality' into their products, or to invent new product categories.

While a few dominant application software providers have grown into large complex organizations such as Microsoft (see Cusamano and Selby 1995) or Intuit, most standard software firms are small entrepreneurial firms with project-based forms of organization. Particularly in their early stages – before a successful product has been launched – standard software firms focus exclusively on product development. Firms race to develop products with the maximum number of features given deadlines imposed by internally announced product launch dates or, more often, innovation races with competitors to launch similar products on the market. To compete, firms generally use a variety of high-powered incentives, such as stock-option schemes and bonuses for meeting development milestones (for ethnographic studies of project organization at standard software start-ups, see Cusamano and Yoffie's (1998) description of the early days of Netscape and Ferguson's (1999) history of web-page authoring software start-up Vemeer). To facilitate the rapid organization of project development teams and reduce career risks caused by the failure of many projects, firms use standardized skill-sets whenever possible. These include industry-wide job descriptions (i.e. programmer, debugger, project organizer, etc.) and technical skills (i.e. language expertise in C++, Java, or Perl).

Collaborative firms

'Collaborative firms' are firms operating in fields with relatively low technological risk but high market risks are organized primarily to deal with appropriability dilemmas (Figure 8.1). If basic technological assets are generic, then, following logic set out by Teece (1986), firms must develop complementary assets that are specific to the firm. While establishing a brand is one possible route, common strategies involve the creation of libraries of technologies that can be linked to the core product and customized extensively for individual clients. Firms can then bundle generic product platforms developed through in-house R&D with complementary investments in customization, implementation, and following on technology consulting. Once customers have purchased a particular firm's technology, lock-in effects often develop due to the sunk costs of purchasing entirely new systems. This can lead to follow-on business as technological upgrades or new services tied in to established platforms are introduced.

From the point of view of managerial risk, co-specialized asset strategies create more complex organizational environments than found in project-based firms focused on more generic R&D. Managers must ask skilled employees to invest in firm-specific technologies, leading to the development of skill-sets that can be difficult to sell on open labour markets. To develop co-specialized assets, extensive team-based work often develops between basic R&D personnel and technicians and consultants involved in customization work for particular clients. A key attribute of a firm's competitive success can be its ability to develop an organizational culture or set of routines enabling different types of professional employees to work well in cross-functional teams. From the point of view of employees, this represents primarily firm-specific and often tacit knowledge that is difficult to sell on the open labour market.

Organizational risks are more complex than those found in project-based firms. Employees must worry about managers pursuing opportunistic employment policies, such as holding wages below industry norms, once extensive firm-specific knowledge investments have been made. Performance incentives may also be difficult to develop, as extensive teamwork across employees with different skill-sets makes it difficult to award individual employee performance (see Miller 1992). While project-based firms revolve around the completion of relatively short-term R&D milestones, the success of collaborative firms is driven by team externalities that develop over multiple development and implementation cycles. Unless managers can assure employees that they will not exploit firm-specific knowledge investments, employees could refuse to make long-term knowledge investments within cross-functional teams, creating patterns of suboptimal work organization that could hurt the performance of the firm.

To manage this constellation of risk, managers must generally create a series of 'credible commitments' (Kreps 1986) not to hold up employees. In game-theoretic terms, their purpose is to transform short-term, single iteration transactions between managers and employees (lasting, for example, one cycle of product development), into repeated games lasting well into the future. Credible commitments often comprise formal rules made by managers, for example to develop a strict code to govern hire-and-fire practices within the firm, or to develop consultative workplace practices or other forms of stakeholder decision-making (see Miller 1992). Reputation-based incentives, for instance regarding norms followed in creating promotion systems, procedures used to award bonuses, or consultative practices between top management and skilled employees regarding major strategy decisions, can also create a long-term equilibrium towards risky skill investments.

Within the software industry, enterprise software segments share market and technological characteristics that encourage the creation of company capabilities similar to collaborative firms. Market segments in this category include enterprise resource planning (ERP), customer relationship management (CRM), groupware, systems integration, e-commerce software providers, and a variety of firms creating sector-specific enterprise tools (e.g. logistics and supply chain management tools). Within enterprise software, network externalities derived by widespread use of particular software platforms across firms are low, limiting the development of 'winner take all' markets often characteristic of standard software. Virtually all enterprise software markets have several established competitors, as it is relatively simple for competitors to develop alternative development paths to introduce similar technologies.

Enterprise software providers use in-house R&D to develop generic software platforms or libraries that are then customized for particular clients. Learning effects created through customization work can, over time, feed back into the overall quality of the firm's generic software platform. Customization and implementation work generates high sunk costs for clients, which can be exploited by enterprise software firms to generate follow-on business such as software upgrades to add new features. This creates more complex organizational structures than those in standard software companies. Successful enterprise software firms must institutionalize teamwork between core developers and teams of software installers and consultants that customize and install the firm's products for particular clients. While standard software firms need to be quick on their feet, enterprise software firms need to develop longer employee commitments to the firm, often entailing investments in firm-specific knowledge that can be risky for staff.

Basic Research Organizations

Technological areas with high market and technology risks encompass the full array of organizational dilemmas facing entrepreneurial ventures. With such technologies it is difficult to use the strategy of developing co-specialized assets to avoid the appropriability issue, as there is little guarantee either that the technology investments will pay off or that a market will exist due to low cumulativeness. This creates high knowledge investment risks for both managers and employees, while also generating competency destruction problems for skilled employees. Technologies with these characteristics tend not to be developed by entrepreneurial technology firms, but are instead pioneered within basic research organizations – primarily universities or pure research labs of very large firms. Within basic research organizations commercial risks may be compensated for through non-commercial goals or, at times, cost-plus governmental financing. Reputational rewards for scientists that make contributions to important basic research puzzles, regardless of their immediate commercial significance, can be used within such organizations. Important basic research contributions within the software industry include research on the organization of data structures, the development of programming languages (e.g. Pascal, C, or the artificial intelligence language LISP), and early research on networking concepts and protocols (e.g. the development of the Ethernet at Zerox Parc). Due to their complexity and the fact that very few entrepreneurial technology firms engage in basic research activities, our subsequent empirical analysis ignores this type of organization.

INSTITUTIONAL FRAMEWORKS AND THE GOVERNANCE OF COMPETENCIES

The software industry is comprised of a number of sub-sectors, each with a different constellation of technological and market risk that combine to create unique organizational challenges. We now explore how national institutional arrangements influence the types of organizational risks that firms can easily govern. To develop this argument, we draw upon typologies of national business systems developed by scholars working within the 'varieties of capitalism' field (Hall and Soskice 2001; Whitley 1999). Based on a dichotomy between 'liberal market economies' or 'LMEs' (the USA, UK, or Canada) and 'coordinated market economies' or 'CMEs' (Germany, Sweden, or Japan), these scholars explain how differences in the historical development of institutional arrangements governing industrial relations, finance, labour markets, and inter-firm relations influence patterns of industrial organization within an economy. Institutional frameworks influence the activities of firms through providing templates or tool-kits firms may use to structure activity. The orientation of these tool-kits advantage the governance of some organizational dilemmas, while impeding others.

Table 8.1 highlights some of the primary institutional differences across CMEs and LMEs. Focusing on Germany and the UK, we explain how contrasting patterns of employment and ownership relations that evolve in

Table 8.1   Institutional framework architectures in Germany and the UK

	CMEs – Germany	LMEs – UK
Labour law	Regulative (coordinated system of wage bargaining; high redundancy costs to laying off employees); bias towards long-term employee careers in companies	Liberal (decentralized wage bargaining; few redundancy costs to laying off employees); few barriers to employee turnover
Company law	Stakeholder system (two-tier board system plus codetermination rights for employees)	Shareholder system (minimal legal constraints on company organization)
Skill formation	Organized apprenticeship system with substantial involvement from industry. Close links between industry and technical universities in designing curriculum and research	No systematized apprenticeship system for vocational skills. Links between most universities and firms almost exclusively limited to R&D activities and R&D personnel
Financial system	Primarily bank-based with close links to stakeholder system of corporate governance; no hostile market for corporate control	Primarily capital-market system, closely linked to market for corporate control and financial ownership and control of firms

relation to these institutions provide incentives and constraints in governing risks associated with different types of entrepreneurial technology firm. This leads to hypotheses pertaining to patterns of comparative institutional advantage across CMEs and LMEs.

Coordinated Market Economies – Germany

Within Germany, patterns of economic organization are organized in nature, primarily due to the embeddedness of large firms within networks of powerful trade and industry associations, as well as a similar, often legally mandated, organization of labour and other interest organizations (for an overview see Katzenstein 1987, 1989). Businesses engage associations to create important non-market collective goods – examples include a strong system of vocational training (Culpepper 2003), extensive inter-firm collaboration over industry-wide technical norms (Tate 2001), and programmes to support the diffusion of new technologies to small- and medium-sized firms (Harding 2001). We now sketch out this system in more detail, emphasizing how German institutions impact on the organization of careers, company organization, and finance within typical companies, as well as the influence these patterns have on the governance used to develop particular innovative competencies.

Turning first to the structure of labour markets, how are careers for scientists and managers organized? In Germany, most employees spend most of their careers within one firm, often after a formal apprenticeship or, in the case of many engineers and scientists, an internship arranged in conjunction with their university degree (Abramson et al. 1997). While there exist no formal laws stipulating long-term employment, redundancy laws create substantial social burdens for firms once they lay off employees. Moreover, German labour has used its power on supervisory boards of large firms as well as its formal consultative rights under codetermination law over personnel policy to obtain unlimited employment contracts (Streeck 1984). As the long-term employment norm for skilled workers was established over the postwar period (see Thelen 1991), it spread to virtually all mid-level managers and technical employees (see Lehrer 1997). A consequence of long-term employment is a dampening of labour markets for mid-career managers and technical employees. 'Hire and fire', though not illegal in Germany, is difficult.

Long-term human resource policies are complemented by Germany's well-known 'stakeholder' model of company organization (Charkham 1995; Edwards and Fischer 1994). Combined with long-term employment, codetermination rights for employees create incentives for management to establish a broad consensus across the firm when major decisions will be made. As unilateral decision-making is limited, it is difficult for German firms to create strong performance incentives for individual employees. Performance rewards tend to be targeted at groups rather than individuals within German firms, and individual performance assessments and bonus schemes are limited.

Until 1999 stock options, one of the most common incentive instruments used in US firms, were very difficult to employ due to a combination of tax and financial regulation limiting the ability of firms to buy and sell their own shares. Though financial reforms have simplified their use, they are still uncommon in Germany and typically, when used below top management, are distributed across large groups of employees to ensure that group rather than individual incentives are maintained. Finally, most career structures are well defined in German firms and based on educational qualifications and seniority within the firm rather than on short-term performance.

This system of company organization tends to 'lock-in' owners, managers, and skilled employees into long-term, organized relationships. Doing so facilitates the creation of credible managerial commitments needed for employees to willingly make firm-specific knowledge investments that are not easily saleable on open labour markets. Human resource policies based on 'competency enhancement' rather than competency destruction are the norm.

Ownership and financial relationships in Germany are strongly influenced by corporate governance rules and a predominately bank-centred financial system. Despite the recent expansion of equity markets, Germany remains a bank-centred financial system. According to 1996 data, while market capitalization as a percentage of gross domestic product was 152 per cent in the UK and 122 per cent in the USA, in Germany it was only 36 per cent (Deutsche Bundesbank 1997). Banks and other large financial actors (e.g. insurance companies) have a strong oversight role on firms through seats on supervisory boards and through continuing ownership or proxy-voting ties with most large German industrial enterprises (see Edwards and Fischer 1994). Most German firms still rely on banks or retained earnings to finance investments. Banks are generally willing to offer long-term financing for capital investments, but not for R&D. German banks usually only offer financing for investments in which collateral exists, for example, fixed investments such as property or long-term capital investments. Banks can adopt a longer-term focus in part because they know that German firms are able to offer long-term commitments to employees and other stakeholders to the firm.

The German system of finance creates difficulties in funding entrepreneurial technology firms for which assets that can be secured against bank loans are often minimal. Moreover, as pointed out by Tylecote and Conesa (1999), banks in 'insider'-dominated corporate governance systems tend to have excellent knowledge of particular firms, but usually do not have the detailed *industry* knowledge that is necessary for investors to channel money into higher-risk technologies. Rather, financing for higher-risk activities is generally provided by venture capitalists, often in conjunction with industry 'angels' that have detailed technical and market expertise within particular industries. Until recently the growth in venture capital has been limited in Germany due to the lack of an investment banking community backed by a stock market that could credibly provide liquidity for initial public offerings. Without such a market, venture capitalists cannot foresee relatively short-term returns on investments that can be used as a refinancing mechanism for future investments or credibly embark on diversification strategies across numerous high-risk investments (see Zider 1999).

While the German system of long-term employment and the stakeholder pattern of company organization have remained stable, in recent years the German financial system has begun to change. Backed in part by government matching subsidies for high-risk investments in new technology sectors, a vibrant venture capital sector has emerged in Germany since the late 1990s (see Lehrer 2000). A new stock market aimed at technology listings, the Neuer Markt, was created in 1998. At its peak it successfully embraced over 300 initial public listings for technology firms and, during 1999 and early 2000 was widely seen as Europe's most successful technology-focused stock exchange. The rapid decline of share prices for technology-related stocks in the USA during the latter half of 2000 and 2001 led to a crisis for the Neuer Markt, which lost over 70 per cent of its value. This led to many firms going bankrupt or being removed from the exchange. Finally, in early 2004 the remaining firms were consolidated into the larger Frankfurt Stock Exchange, effectively closing the Neuer Markt. Nevertheless, the existence of a vibrant venture capital community and the Neuer Makrt may herald an important shift in the viability of entrepreneurial business models in Germany. One of the important issues investigated in our empirical analysis is whether the corporate governance of software companies listed on the Neuer Markt during its peak in 2000 resonates with long-established trends in Germany or represents a fundamental shift towards liberal market patterns.

Liberal Market Economies – UK

The UK is characterized by an LME. Public policy within the UK is more 'neutral' in character, often imposing strong regulation on the characteristic of particular markets but rarely privileging particular market participants over others. The influence of organized associations on government policy, whether representing business or labour, is generally weak, with few statutory or implicit rights over the nature of business regulation. Industry and trade associations in the UK engage primarily in lobbying and business promotion; compared to CMEs their role in organizing non-market patterns of coordination across firms is weak. As a result, market forms of organization govern much of the economy, and the role of industry in education and vocational training is limited. Business organization depends primarily on market transactions and the use of a flexible, enabling private legal system to facilitate a variety of complex contracting arrangements. Compared to coordinated market economies, high-powered performance incentives are more readily available to align interests within and across organizations (see Easterbrook and Fischel 1991).

Institutional frameworks in the UK encourage few, if any of the company organizational and financial structures needed to pursue long-term incremental innovation strategies associated with collaborative firms, but are ideally suited to the competitive requirements for radical innovation within project-based firms. Particularly after the Thatcher era in the early 1980s, labour markets are deregulated in the UK and collective bargaining has become decentralized. Bargaining over the level and forms of remuneration for most professionals and managers occurs on a company by company basis, and often between individuals and the firm. Most firms offer limited employment contracts, poaching is widespread and an extensive 'headhunting' industry has emerged alongside most regional agglomerations of high-technology firms in Cambridge and elsewhere. This allows firms to quickly build or shed competencies as they move in and out of different technology markets. The existence of deep external labour markets increases the viability of 'competency destroying' human resource strategies, as highly skilled employees can commit to a high-risk job with the knowledge that large labour markets exist for a variety of standardized skill-sets within technology sectors.

Compared to the 'social' construction of German firms, the property rights structure of most UK firms is financial in nature (see Roe 1993). No legally stipulated codetermination rights for employees or other stakeholders exist. This allows owners to create high-powered incentive structures for top management (i.e. very high salaries often paid in company shares or share options), who are then given large discretion in shaping organizational structures within the firm. The top management of most UK technology firms attempt to create similarly high-powered incentive structures for valued employees within the firm. These structures include the use of performance bonuses as a high percentage of overall remuneration, opportunities for star performers to quickly advance through the firm, and much unilateral decision control (see Lehrer and Darbishire 2000 for a case study of the development of this system within a large UK organization, British Airways).

A similar German–UK contrast holds concerning finance. Large capital markets in the UK create viable exit options for venture capital. As a result the UK has long had Europe's most developed venture capital and investment banking markets. This financing tends to be short-term in nature, meaning that funds will dry up if firms fail to meet development goals or if products fail to live up to expectations in the marketplace. However, so long as the possibilities for high, often multiple returns on investment exist, a large market of venture capitalists and, at later stages of company development, more remote portfolio investors stand ready to invest in technology firms. The broad institutional structure of the UK largely explains why this is the case. First, given the deregulated nature of labour markets, high-quality managers and scientists can be found to fuel the growth of highly successful firms. Second, investors know that performance incentives can be managerially designed to 'align' the risk/return preferences of investors with rewards for top management and employees of particular firms.

Hypotheses

Within CMEs such as Germany a preponderance of long-term employment, an emphasis within most firms on predictable patterns of career advancement, and norms encouraging relatively low-powered and collective systems of performance incentives strongly advantage the creation of consensual workplace relations within firms and systematic investments in firm-specific knowledge across employees needed to sustain collaborative firms such as those found in enterprise software. On the other hand, German institutional arrangements appear less suited to developing competencies associated with project-based firms. It is difficult for German firms to quickly move in and out of markets characterized by rapidly evolving technologies. Since most employment contracts are unlimited, top managers of German firms must think twice before creating new competencies in high-risk areas, for cutting assets is difficult. Similarly, it is difficult for German firms to create the high-powered performance incentives.

This summary leads to our first two hypotheses:

1. CMEs have a comparative institutional advantage in creating collaborative firms but
2. Have a comparative institutional disadvantage in the governance of project-based firms.

Within LMEs such as the UK, the development of individualized employment relationships and flexible patterns of company organization within large firms strongly impacts on the development of project-based entrepreneurial firms. Hire-and-fire, when embraced by most companies within a sector, can be used to create large external labour markets for most skills, easily tipping firms into patterns of labour market coordination favouring highly flexible human resource policies needed to face technical and market risks. Financial markets support the development of venture capital, the terms of which can be credibly linked to the completion of performance milestones within project-based firms. On the other hand, employees facing this pattern of labour market and company organization will be extremely reluctant to develop patterns of firm-specific skill development needed to support collaborative entrepreneurial firm strategies, while the short-term orientation of most venture capital investments will impede their corporate governance.

This leads to two further hypotheses:

3. LMEs have a comparative institutional advantage in the governance of project based firms, but
4. Have a comparative institutional disadvantage in the governance of collaborative firms.

While not discussed in the preceding analysis, the typology of entrepreneurial technology firms also included partnerships, for which no particular patterns of comparative institutional advantage exist. This leads to a final hypothesis:

5. Neither LMEs nor CMEs have comparative institutional advantages or disadvantages in the governance of partnerships.

EMPIRICAL ANALYSIS: SUB-SECTOR SPECIALIZATION WITHIN THE UK AND GERMAN SOFTWARE INDUSTRIES

The statistical technique of cluster analysis is well suited to measure whether there are distinct types of entrepreneurial technology firms and the extent to which the distribution of these types across countries differs. Cluster analysis is a technique which helps identify a discrete number of groups or classes whose members (in this case firms) tend to share similar values across a number of dimensions. We employ a cluster analysis to examine whether groups of software firms exist with characteristics resembling different types of entrepreneurial technology firms. If firms cluster into stable groupings that represent distinct sub-sectors within the software industry *and* if the distribution of nationality across clusters is significantly different, then this will help confirm our hypotheses linking national institutional factors to patterns of sub-sectoral industry specialization in the software industry.

Sample

To test our hypotheses we gathered data on 193 German and UK software companies active during the year 2000. The German companies were listed on the Neuer Markt. The UK software companies are included in Techmark, a special index on the London Stock Exchange for high-technology companies. Our sample includes all German and UK software companies on those exchanges for which comprehensive data were readily available; foreign software companies were excluded.[1] After removing three outliers whose values were highly skewed (e.g. negative shareholders' equity due to insolvency) we were left with 190 companies roughly equally represented by nationality (98 British and 92 German).

Measures

Two types of variables were used in the analysis: 'structural variables', which measure institution-related features of companies and are used to generate clusters, and 'auxiliary variables', which capture a number of supplementary characteristics which are used to validate the analysis. Table 8.2 summarizes the variables used in the analysis and data sources for these variables. Unless otherwise mentioned, all company data are for the business year 2000.

Table 8.2   Variable names, descriptions and data sources

Variable name	Variable description	Data source
Structural variables		
Ownership concentration	Percentage of shares owned by founder, founder's family and top managers	Germany – data file from Deutsche Börse UK – Company annual reports
Stock options	Stock options granted to managers and employees and not yet exercised, as a percentage of total shares issued	Company annual reports, stock exchange web sites
Bank debt	Bank debt as a percentage of the total assets of the company	Company annual reports
Shareholder equity	Stockholders' equity as a percentage of total assets	Company annual reports
Software specialization index	A special measure including measures of the business model alignment, process architecture and organizational structure of the companies	Company web sites, IPO prospectuses and company annual reports
Auxiliary variables		
Sales	Total annual company sales	Company annual report, or IPO prospectus for recently listed companies
Employment	Total number of company employees	Company annual report, or IPO prospectus for recently listed companies

Table 8.2   (Continued)

Variable name	Variable description	Data source
Age	2000 minus founding year of the company	Company web sites, IPO prospectuses and company annual reports
Sales growth	Percentage increase in total annual company sales from 1999–2000	Company annual report, or IPO prospectus for recently listed companies
Return on assets	Operating results of company before interest and taxes (EBIT) divided by total company assets	Company annual report, or IPO prospectus for recently listed companies
Return on investment	Operating results of company before interest and taxes (EBIT) divided by total shareholder equity	Company annual report, or IPO prospectus for recently listed companies
Sales margin	Operating results of company before interest and taxes (EBIT) divided by total annual company sales	Company annual reports, IPO prospectuses for recently listed companies
R&D as a percentage of sales	Research and development expenditures as a percentage of total sales	Company annual reports, IPO prospectuses for recently listed companies
R&D as a percentage of costs	Research and development expenditures as a percentage of total costs	Company annual reports, IPO prospectuses for recently listed companies

Cluster Generation

Our first step was to identify a set of variables measuring structural characteristics of UK and German software firms. These variables were defined deductively on the basis of varieties of capitalism theory. We selected five structural characteristics of software firms, the values of which should be influenced by finance, corporate governance, and human resource institutions in CMEs and LMEs. We discuss the expected values of each of these structural variables across CMEs and LMEs and across different types of entrepreneurial technology firms below. We used these variables to generate clusters based on the premise that differences in national institutional structures across the UK and Germany should be reflected in variation in these structural characteristics across firms.

For four of the variables used (ownership concentration, use of stock options, bank debt, and shareholder equity) standard measures were

Table 8.3   Composition of the Standardized Software Index (SSI)

Indicator	Value
Does the company sell its own software?	0.1
Does the company focus on large, medium or small-sized customers?	0.1
Is the process itself that the software is applied to, standardized?	0.1
Does the company have a large marketing and sales division?	0.1
Does the company participate in a sales partner (reselling) program?	0.1
Does the company have a software development program with an OEM?	0.1
Was the product developed in an independent development environment?	0.1
Does the company offer consulting or implementation services?	0.1
Is third-party software also offered by the company?	0.1
Is maintenance part of the sales agreement?	0.1
Sum (SSI)	1.0

available. To more accurately measure the organization of a firm's R&D processes, we used a novel index (standardized software index (SSI)) to measure the relative standardization of a company's products.[2] The measure we used is based on 10 indicators or aspects of standardization, listed in Table 8.3. A maximum score of 0.1 is possible on each of the 10 factors, thus resulting in a lowest possible score of 0 and highest possible score of 1.

This measure is intended to measure the following qualitative dimensions of standardization within the firm:

- Importance of sales from standardized (as opposed to project-oriented) company divisions;
- Sales growth dynamic of standardized company divisions;
- Proportion of sales of own software (as opposed to services and third-party software);
- Relationship of technical personnel to sales and marketing personnel;
- Ratio of own products to number of customers.

Before evaluating whether cluster characteristics for each country correspond to our predictions, we first examine how each of the five firm characteristics reflect corporate governance, finance, or human resource institutions associated with coordinated and liberal market economies.

1.   Ownership concentration

This variable measures the percentage of shares owned by the top management of the company and is most closely linked with the national corporate governance system. Concentrated 'insider' ownership is a fundamental characteristic of corporate governance within CMEs such as Germany, particularly among small- and medium-sized firms. This contrasts with the 'Silicon Valley' model of small firms frequently found in LMEs such as the USA and the UK, where majority ownership by one or more venture capitalists before an IPO and a rapid increase in dispersed ownership after the IPO are typical. A high score on this variable would therefore be expected for German firms and a lower score for UK firms as a reflection of institutional differences.

2. Stock options
This variable measures the amount of stock options granted to managers and employees as a percentage of the total stock of a company outstanding. It is most closely linked to the characteristics of institutions influencing human resource practices (i.e. labour markets and company law). Generous use of stock options to motivate managers and employees are associated with LMEs, which have higher-powered incentive structures and more individualistic performance incentives. CMEs in contrast tend to have more collective systems of remuneration aimed at rewarding qualifications gained and seniority rather than short-term performance. LMEs are thus associated with a high score on this variable and CMEs with a low score.

3. Bank debt
This variable describes the volume of loans from banks outstanding as a percentage of the total balance sheet. It is most closely linked to the characteristics of the financial system, particularly to the significance of banks. Bank-based financing is a predominant characteristic of CMEs, which tend to have bank-dominated financial systems, whereas companies in LMEs tend to rely less on bank debt than on market-based forms of finance.

4. Shareholder equity
This variable measures the ratio of shareholder equity to total balance sheet assets. It is most closely related to the financial preferences of investors, and is influenced by financial system and corporate governance institutions. Since shareholder equity is the financial reserve that companies draw upon in case of financial losses, this ratio should reflect the risk–reward trade-off that shareholders face. This ratio should be higher in the case where companies are riskier. Low ratios in contrast are needed where expected returns are lower but less variable (i.e. less risky). In capital-market-based systems investors should demand relatively high shareholder equity, as it is generally more difficult to monitor the behaviour of companies. In a bank-based system banks can better monitor companies. A higher score on this is variable is thus associated with LMEs and a low score with CMEs.

5.   Software Standardization Index (SSI)

As discussed earlier, SSI is a unique measure associated with the human resource structure of the firm. The relationship between the level of standardization and institutional frameworks is more complex. Returning to our typology of entrepreneurial technology firms, we expect standardization to be very high within project-oriented standard software firms and very low within IT service firms. Enterprise software firms develop collaborative organizations that integrate a substantial amount of generic R&D with complementary assets used to customize software for clients, leading to intermediate levels of standardization. As there are no organizational dilemmas associated with very high levels of customization as found in partnerships, both LMEs and CMEs should be able to accommodate clusters of firms with very low levels of software standardization. Our theoretical framework predicts that LMEs can best organize project-based organizations used to generate innovations in areas with high technological risk, which should lead to high SSI scores. Finally, we predict that CMEs can best govern collaborative enterprises in which co-specialized R&D and customization assets must develop. SSI can thus also be of an intermediate value for CMEs.

Table 8.4 summarizes this discussion of structural variables and their expected values in CMEs and LMEs.

An important issue in the use of cluster analysis is the problem of multicollinearity, which could lead to the over- or underweighting of one or more variables in the estimation of cluster means. Pearson correlation coefficients can be examined to see how highly variables used to estimate cluster means (i.e. the position of clusters) are correlated with each other. A value close to 1 would indicate high correlation and thus would pose a problem for the analysis.

To test for the possibility of multicollinearity correlation coefficients between the structural variables were calculated. None of the 'structural' variables used to estimate the cluster means had a correlation coefficient of higher than 0.6. This indicates that these five variables are sufficiently independent of one another to allow use in generating clusters.

An additional issue is whether variables should be standardized. The variation along different variables may differ substantially, i.e. values along variable X might vary between 50 and 60 on a scale of 0–100 versus variance in along variable Y from 20 to 80 along the same scale. Standardization manipulates the values so that the distance from low to high values along different variables are rendered comparable. Standardization however may lead to different weightings of the variables. This can be dealt with by seeing if the results change substantially when using non-standardized variables instead. When doing this we found no substantial difference, and thus

Table 8.4   Structural variables used and their expected values

	Definition	Institutional influences	CME expected value	LME expected value
Ownership concentration	Ownership concentration	Corporate governance	High	Low
Stock options	Outstanding stock options as a percentage of total number of outstanding shares	Human resource organization/ corporate governance	Low	High
Bank debt	Percentage of total assets (or liabilities plus shareholder equity) held as bank debt	Corporate governance	High	Low
Shareholder equity	Shareholder equity/ balance sheet total assets	Corporate governance	Low	High
Standardization index	The relative standardization of a company's products (an index)	Human resource organization	Medium (but low is also possible)	High (but low is also possible)

elected to use standardized variables due to the ease of presentation of results.

Finally, we needed to decide which algorithms or specific types of cluster analysis should be used to estimate the number of clusters and the cluster means. Following the recommendations of Ketchen and Shook (1996), a combination of hierarchical and non-hierarchical algorithms were used in estimating the number and values of the software clusters. First, a hierarchical algorithm using Ward's method was used to identify outliers. After outliers were removed, the analysis was repeated in order to identify the cluster centroids (mean values of variables) and the optimal number of clusters. This analysis led to our adoption of a four-cluster solution, due a large jump from the third relative to the fourth solution and a relatively small increment between the four-cluster and the five-cluster solution. As a next step, a non-hierarchical method (using K-means) was applied to the centroid means estimated by the Ward method for the best cluster solution. The non-hierarchical method allows for iterations in estimating final cluster means and membership in order to arrive at an optimal solution. Iterations

Table 8.5   Standardized cluster means (structural variables)

	1	2	3	4
Ownership concentration	−0.92313	−0.59606	0.87246	0.39461
Stock options	0.06386	0.70214	−0.53209	−0.11148
Bank debt	−0.35820	−0.42114	−0.31351	1.86547
Shareholder equity	−0.10750	0.39619	0.45286	−1.47046
Standardization index	−1.00114	0.99074	0.02201	−0.16172
Number of firms (n=190)	46	51	63	30

Table 8.6   National distribution by cluster*

	1	2	3	4	Total
UK	**39 (84.7%)**	**39 (76.5%)**	10 (15.7%)	10 (33.3%)	98
Germany	7 (15.3%)	12 (25.5%)	**53 (84.3%)**	**20 (66.7%)**	92
Total	46	51	63	30	190

Note: *Bold signifies main clusters.

are performed until no company switches cluster membership. The results for the four-cluster solution are displayed in Table 8.5.

ANALYSIS

We use Table 8.5 to examine whether there is a relationship between national institutional frameworks and the characteristics of firms. We start by assessing whether cluster membership is associated with firms of particular nationality. If not, then we could still assess whether these clusters share characteristics of particular types of entrepreneurial technology firms, but the idea that national institutional frameworks drive patterns of sub-sector specialization would be invalidated.

Table 8.6 breaks down cluster membership by nationality, and shows that cluster membership is strongly correlated with nationality. Clusters 1 and 2 are predominantly British whereas clusters 3 and 4 are mainly German.

While membership in these clusters is strongly correlated with particular nationalities, this does not necessarily show that the characteristics of firms within particular clusters match those associated with coordinated and liberal market economies. As discussed earlier, this cluster was generated through data associated with five firm-level characteristics that are strongly associated with corporate governance, finance, and human

*Table 8.7   Predicted institutional values and stylized cluster results**

	Clusters 1 UK – testing for LME values	Clusters 2 UK – testing for LME values	Clusters 3 Germany – testing for CME values	Cluster 4 Germany – testing for CME values
Ownership concentration				
Predicted value	Low	Low	High	High
Result	**Very low**	Low	High	High
Stock options				
Predicted value	High	High	Low	Low
Result	*Neutral*	**Very High**	**Very Low**	*Neutral*
Bank debt				
Predicted value	Low	Low	High	High
Result	Low	**Very low**	*Low*	**Very High**
Shareholder equity				
Predicted value	High	High	Low	Low
Result	*Neutral*	High	*High*	**Very Low**
Standardization index				
Predicted value	Low or High	Low or High	Low or Neutral	Low or Neutral
Result	**Very Low**	**Very High**	**Neutral**	Low

Notes:
*Bold signifies result strongly supports predicted value; italics signify result does not support predicted value. Predicted values are based on Table 8.4.
Cluster results are based on Table 8.5 using the following classification scheme:
Very high = standardized cluster mean scores <0.5; high = 0.15 <score <0.5; 0 = neutral (−0.15 <score <−0.15); low = (−0.15 <score <−0.5); very low = (score<−0.5).

resource institutions across CMEs and LMEs. We now evaluate whether cluster characteristics for each 'national' cluster correspond to the correct institutional predictions.

Drawing upon the institutional predictions in Table 8.4 and the standardized cluster mean values in Table 8.5, Table 8.7 helps evaluate whether cluster characteristics for each 'national' cluster correspond to the correct institutional predictions.

These results support the claim that institutional frameworks shape the characteristics of firms in the two countries. Of the 25 predictions, 20, or 80 per cent, are supported. Seven of these values (listed in bold) are strongly supported, meaning that relatively high or low standardized mean values have been generated (i.e. mean value scores were higher than 0.5 or lower than −0.5), while the standardization index for cluster 3 also closely matches a predicted result of 'neutral' for CMEs. Of the unconfirmed predictions, three were 'neutral' results that were relatively close to the 'high' or 'low' predicted value.

Only two predictions are far off mark – bank debt and shareholder equity within cluster 3. This result is surprising, as this cluster represents 63 firms, 84 per cent of which are German. Both bank debt and shareholder equity are strongly related to patterns of financing. High shareholder equity and low bank debt are dominant characteristics of equity-leveraged financing through capital markets – a pattern closely associated with LMEs. Evidence that a large group of German firms have adopted capital-market-based financing methods suggests that recent reforms to improve high-risk financing in Germany might be having a significant effect. Corporate governance, however, might still conform to 'insider' dominated practices, as ownership concentration is high for cluster 3. However, notice also that patterns of financing in the other predominately 'German' group of firms, cluster 4, has very strong standardized mean scores for traditional patterns of financing with CMEs, very high use of debt financing and very low shareholder equity. This evidence supports claims that the German financial system is changing. Patterns of financing and corporate governance within the UK, however, appear for the most part to confirm to the expected liberal market model. The only surprising result is the 'neutral' result for shareholder equity in cluster 1.

Human resource patterns also generally conform to predictions. There are no incorrect results for the standardization index, and standardized mean scores for three of the sectors were relatively extreme. This offsets the weaker predictive value for this variable (as two of a possible three results are theoretically allowed). The stock option scores are strongly supportive for clusters 2 (very high for the UK) and 3 (very low for Germany) but are only neutral for clusters 1 and 4. Examining the data more closely, the result for cluster 4 (standardized mean score −0.11) is very close to our threshold score for a low use of stock options. The score for cluster 1 (0.06) also has the correct sign, but given the high population of UK firms in this group (39) the result shows that not all UK entrepreneurial technology firms strongly rely on stock options as a human resource management tool.

In sum, Table 8.7 helps confirm the core premise of comparative institutional analysis – namely that the orientation of institutional frameworks shape the characteristics of firms. However, our results also show that firms have substantial leeway in adopting organizational structures to match competitive dynamics within particular market segments. A crucial issue in this regard is the degree to which firms engage institutions to shape patterns of competitive advantage or, if national institutions are not useful, seek to develop managerial practices that avoid them. Our earlier theoretical analysis and hypotheses developed the idea of comparative institutional advantage. This concept suggests that institutions do more than shape firm characteristics, they provide useful tool-kits that help (or hinder) the governance of the core organizational dilemmas facing particular types of firm.

CONCLUSION

This chapter has developed and tested a theoretical framework to evaluate the impact of national institutional frameworks on the organization and innovation strategy of entrepreneurial technology firms. The chapter has created a typology of different types of entrepreneurial technology firms. Drawing from research on sectoral systems of innovation and from the economics of organization, we examined how differences in the market and technological organization of different sub-sectors of new technology industries such as software create different constellations of organizational risk facing managers of entrepreneurial technology firms. An important insight drawn from this typology is that high-technology industries such as software are comprised of several *sub-sectors*, for which underlying market, technological, and organizational dynamics differ. Our empirical analysis demonstrated that firms in the software industry cluster into distinct industry groups – supporting the theory underlying our typology, as well as broader research in strategy management on the heterogeneity of firms.

Drawing on our typology of sub-sectoral patterns of innovation, we have examined the relationship between national institutional frameworks across different 'models of capitalism' and the development of particular organizational competencies. This analysis predicts that LMEs such as the UK enjoy comparative institutional advantage in project-oriented firms, but face difficulties in sustaining collaborative enterprises, while the opposite trade-offs occur within CMEs such as Germany. This analysis complements a recent drive to create more 'firm-centred' institutional analyses (Hall and Soskice 2001). Our empirical evidence again confirms our central hypotheses concerning patterns of comparative institutional advantage within the software industry. As much previous institutional theory predicts, UK firms do enjoy advantages in governing project-based firms focused on 'radical innovations', while German firms are advantaged in establishing more enduring collaborative enterprises focused on the generation of complementary assets to generic core technologies. Evidence was also consistent, though weaker, with the idea that partnership models of organization are institutionally 'neutral' in terms of governance.

One contribution of this analysis is to show that European economies can perform well in emerging technology industries such as biotechnology and software. These economies do so, however, not by radically altering institutional frameworks to mimic the US liberal economy model, but by seeking out sub-segments within these segments in which firms can embrace long-standing comparative institutional advantage. Evidence presented in this chapter has documented the existence of important sub-sectors, such as enterprise software, in which patterns of company organization and related business strategy need to develop complex organizational structures focused on 'competency-enhancing' human resource management. Firms within coordinated market economies such as Germany have specialized in these technologies not as a 'second-best' solution, but because the institutional organization of these business systems create institutional advantages in resolving the managerial dilemmas that characterize these sub-sectors.

An implication of this analysis is that trade-offs exist in terms of designing institutions that foster entrepreneurial technology firms. Because different types of technology firms differ in their core organization, their optimal governance requires their embeddedness in different innovation systems. Thus, while the USA has a large lead in fostering new technology firms, as key technological drivers diffuse through the international economy, one can expect that a division of labour will emerge cross-nationally. While institutions associated with the USA (and the UK) innovation systems conducive to success in business models demanding extreme flexibility (and competency destruction), German and other 'organized' economies might promote superior innovation dynamics in areas dominated more by business integration and appropriability risks.

The focus on sub-sectors also sheds light on the organization of more 'radically innovative' technological segments such as standard software. Our analysis resonates with a number of important studies of the institutional organization of high-technology regions such as Silicon Valley (Kenney 2000; Almeida and Kogut 1999; Saxenian 1994). We share with these studies the suggestion that low technological cumulativeness and resulting 'competency destruction' across clusters of new technology start-ups can be best managed by the creation of extremely fluid labour markets within regional economies. While most studies of Silicon Valley and related technology clusters have a regional focus, we focus primarily on broader national institutional frameworks that structure patterns of coordination across particular sectors and regions within the economy. Doing so helps explain broad differences in technological specialization across economies, but cannot explain the relatively rare development of regional economies capable of fostering high levels of technological intensity across start-up firms within particular economies.

In other words, there are more 'degrees of freedom' between the orientation of national institutional frameworks and the ability of managers across groups of firms to develop innovative competencies than is suggested by varieties of capitalism theory. Linking insights from varieties of capitalism research with the emerging literature on regional technology clusters is an important area for future research. A firm-centred approach, focused on relatively stable constellations of organizational and strategic dilemmas faced by managers across particular sub-sectors, can help push this agenda forward.

ACKNOWLEDGEMENT

For extensive research assistance for this chapter we are grateful to Lutz Engelhardt and Jana Meier. We gratefully acknowledge funding from the European Commission through the TSER program.

NOTES

1. Only two software companies, SAP AG and Software AG, are listed on the main segment of the Frankfurt stock exchange. Some UK software companies are also listed on AIM (Alternative Investment Market); however, these were excluded due to the poor quality of data available on AIM companies.
2. We acknowledge here the invaluable contributions of our research assistant, Lutz Engelhardt, in developing this index.

REFERENCES

Abramson, H., N. Proctor, P. Reid, U. Schmoch and J. Encarnacao (eds) (1997), *Technology Transfer Systems in the United States and Germany: Lessons and Perspectives*, Washington, DC: National Academy Press.

Almeida, P. and B. Kogut (1999), 'Localization of knowledge and the mobility of engineers in regional networks', *Management Science*, **45** (7), 905–18.

Bahrami, H. and S. Evans (1995), 'Flexible re-cycling and high-technology entrepreneurship', *California Management Review*, **37** (3), 62–88.

Breschi, S. and F. Malerba (1997), 'Sectoral innovation systems: technological regimes, Schumpeterian dynamics, and spatial boundaries', in C. Edquist (ed.), *Systems of Innovation: Technologies, Institutions and Organizations*, London: Pinter, pp. 130–55.

Casper, S., M. Lehrer and D. Soskice (1999), 'Can high-technology industries prosper in Germany: institutional frameworks and the evolution of the German software and biotechnology industries,' *Industry and Innovation*, **6** (1) 5–24.

Charkham, J. (1995), *Keeping Good Company: A Study of Corporate Governance in Five Countries*, Oxford: Oxford University Press.

Culpepper, P. (2003), *Creating Cooperation: How States Develop Human Capital in Europe*, New York: Cornell University Press.

Cusamano, M. And R.W. Selby (1995), *Microsoft Secrets: How the World's Most Powerful Software Company Creates Technology, Shapes Markets, and Manages People*, New York: Free Press.

Cusumano, M. and D. Yoffie (1998), *Competing on Internet Time: Lessons from Netscape and its Battle with Microsoft*, New York: Free Press.

Deutsche Bundesbank (1998), *Monthly Report*, April.

Easterbrook, F. and D. Fischel (1991), *The Economic Structure of Corporate Law*, Cambridge, MA: Harvard University Press.

Edwards, J. and K. Fischer (1994), *Banks, Finance and Investment in Germany*, Cambridge: Cambridge University Press.

Ferguson, C. (1999), *High Stakes, No Prisoners: A Winner's Tale of Greed and Glory in the Internet Wars*, New York: Crown Business.

Hall, P. and D. Soskice (2001), 'An introduction to varieties of capitalism', in P. Hall and D. Soskice (eds), *Varieties of Capitalism: The Institutional Foundations of Comparative Advantage*, Oxford: Oxford University Press, pp. 1–70.

Harding, R. (2001), 'Competition and collaboration in German technology transfer', *Industrial and Corporate Change*, **10** (2), 389–417.

Hollingsworth, R. (1997), 'Continuities and changes in social systems of production: the cases of Germany, Japan, and the United States', in R. Hollingsworth and R. Boyer (eds), *Contemporary Capitalism: The Embeddedness of Institutions*, Cambridge: Cambridge University Press, pp. 265–310.

Katzenstein, P. (1987), *Policy and Politics in West Germany: Towards the Growth of a Semi-Sovereign State*, Philadelphia: Temple University Press.

Katzenstein, P. (1989), 'Stability and change in the emerging Third Republic', in P. Katzenstein (ed.), *Industry and Politics in West Germany: Toward the Third Republic*, Ithaca, NY: Cornell University Press, pp. 307–53.

Kenney, M. (ed.) (2000), *Understanding Silicon Valley: The Anatomy of an Entreprenurial Region*, Stanford, CA: Stanford University Press.

Ketchen, D. and C. Shook (1996), 'The application of cluster analysis in strategic management research: an analysis and critique', *Strategic Management Journal*, **17** (6), 441–58.

Kreps, D. (1986), 'Corporate culture and economic theory', in J. Alt and K. Shepsle (eds), *Rational Perspectives on Political Science*, Cambridge: Cambridge University Press, pp. 90–143.

Lehrer, M. (1997), 'German industrial strategy in turbulence: corporate governance and managerial hierarchies in Lufthansa', *Industry and Innovation*, **4** (1) 115–40.

Lehrer, M. (2000), 'Has Germany finally solved its high-tech problem? The recent boom in German technology-based entrepreneurship', *California Management Review*, **42** (4), 89–107.

Lehrer, M. and O. Darbishire (2000), 'Comparative managerial learning in Germany and Britain: techno-organizational innovation in network industries', in S. Quack, G. Morgan and R. Whitley (eds), *National Capitalisms, Global Competition and Economic Performance*, Amsterdam: John Benjamins, pp. 79–104.

Malerba, F. and L. Orsenigo (1993), 'Technological regimes and firm behavior', *Industrial and Corporate Change*, **2** (1), 45–71.

Milgrom, P. and J. Roberts (1993), *Economics, Organization, and Management*, Englewood Cliffs, NJ: Prentice Hall.

Miller, G. (1992), *Managerial Dilemmas*, Cambridge: Cambridge University Press.

Mowery, D. (1999), 'The computer software industry', in D. Mowery and R. Nelson (eds), *Sources of Industrial Leadership: Studies of Seven Industries*, Cambridge: Cambridge University Press, pp. 133–68.

Mowery, D. and R. Nelson (eds) (1999), *Sources of Industrial Leadership: Studies of Seven Industries*, Cambridge: Cambridge University Press.

Perrow, C. (1985), *Normal Accidents: Living with High-risk Technologies*, Princeton, NJ: Princeton University Press.

Powell, W. (1996), 'Inter-organizational collaboration in the biotechnology industry', *Journal of Institutional and Theoretical Economics*, **152** (1), 197–215.

Roe, M. (1993), 'Some differences in corporate structure in Germany, Japan and the US', *Yale Law Journal*, **102**, 1927–2003.

Saxenian, A. (1994), *Regional Advantage: Culture and Competition in Silicon Valley and Route 128*, Cambridge, MA: Harvard University Press.

Shapiro, C. and H. Varian (1999), *Information Rules: A Strategic Guide to the Network Economy*, Cambridge, MA: Harvard Business School Press.

Streeck, W. (1984), *Industrial Relations in West Germany: A Case Study of the Car Industry*, New York: St. Martin's Press.

Streeck, W. (1992), 'On the institutional preconditions of diversified quality production', in W. Streeck (ed.), *Social Institutions and Economic Performance: Studies of Industrial Relations in Advanced Capitalist Economies*, London and Newbuy Park, CA: Sage, pp. 1–40.

Tate, J. (2001), 'National varieties of standardization', in P. Hall and D. Soskice (eds), *Varieties of Capitalism: The Institutional Foundations of Comparative Advantage*, Oxford: Oxford University Press, pp. 442–73.

Teece, D. (1986), 'Profiting from technological innovation: implications for integration, collaboration, licensing, and public policy', *Research Policy*, **15** (6), 285–305.

Thelen, K. (1991), *Union of Parts: Labor Politics in Post-war Germany*, Ithaca, NY: Cornell University Press.

Tylecote, A. and E. Conesa (1999), 'Corporate governance, innovation systems, and industrial performance', *Industry and Innovation*, **6** (1), 25–50.

Utterback, J.M. (1996), *Mastering the Dynamics of Innovation*, Boston, MA: Harvard Business School.

Whitley, R. (1999), *Divergent Capitalisms: The Social Structuring and Change of Business Systems*, Oxford: Oxford University Press.

Woodward, J. (1965), *Industrial Organization: Theory and Practice*, Oxford: Oxford University Press.

Zider, B. (1998), 'How venture capital works', *Harvard Business Review*, **76** (6), 131–9.

9. The globalization of management consultancy firms: constraints and limitations

Glenn Morgan, Andrew Sturdy and Sigrid Quack

INTRODUCTION

This chapter is aimed at providing a framework for the analysis of organizational structures and processes in the global management consulting industry. Our basic question is, why do global consulting firms exist? What distinctive advantages (if any) are they able to bring to their clients and the consulting task which cannot be achieved by 'national' firms? Consideration of this question leads us into alternative modes of internationalization in this sector. Economists in the field of international business have long posed this question in relation to manufacturing firms (see for example Dunning 1993). However, their answers tend to be limited to economic considerations and ignore the ways in which issues of organizational structure, power and processes impact on the internationalizing strategies of firms. More recently other authors have posed the same question specifically in relation to professional services firms (Aharonhi 2000; Lowendahl 2000; Nachum 2000; Roberts 1998, 1999, 2004). These authors have argued that there are specific characteristics of professional services that require an adaptation of the dominant models of internationalization. These relate to the distinctive interface between clients and suppliers in these contexts where co-presence and interaction is typically essential. This interaction in conditions where knowledge is ambiguous and/or clients may be less 'knowledgeable' than the professionals about the nature and quality of the services delivered has tended also to lead to national regulatory regimes controlling how some professional services are delivered, monitored and controlled. These factors have tended to militate against the globalization of professional services. However, in recent years, the development of professional services outside strong regulatory frameworks of practice (such as in the case of management consultancies and advertising agencies), the gradual decline of national regulatory regimes under pressures of 'free trade' and the increased international standardization of certain forms of professional services (such as audits) has opened up more possibilities for the internationalization of firms in this area. The general phenomenon of what Giddens (1990) has labelled 'time–space distanciation' in theory makes it easier both to maintain communication and control across widely spread national contexts and to facilitate forms of cross-national team-building and cooperation. For all these reasons, the services sector in general and professional services in particular has seen a massive expansion of international activity over the last two decades. Bryson et al., for example, state that 'the value of world commercial services exports has increased some 3.5 times between 1980 and 1999' (Bryson et al. 2004, p. 217). The 2004 World Investment Report from UNCTAD noted that 'on average, services accounted for two-thirds of total FDI inflows during 2001–2, valued at some $500 billion' (UNCTAD 2004, p. *xx*). Interestingly, the report went on to state that 'as the transnationalization of the services sector in home and host countries lags behind that of manufacturing, there is scope for a further shift towards services' (ibid.). This suggests that the organizational issues concerned with the internationalization of service firms have not disappeared. It is these organizational issues which lie at the heart of our concerns.

In order to assess these organizational issues, we refer to the emerging literature which goes beyond the issue of the economic advantages of internationalization to multinational corporations (MNCs) towards considering how organizations can manage the complex interrelationships between the different parts of multinationals (Morgan et al. 2001; Geppert et al. 2003; Kristensen and Zeitlin 2005; Whitley 2005). What organizational forms enable multinationals to achieve these proposed advantages and what are the limitations to this process? Most of this discussion has taken place in the context of the manufacturing industry (cf. Morgan and Quack 2005b) and part of our objective is to show how a similar analysis can be made relevant particularly for management consultancy firms.

The chapter consists of the following sections. First, we begin by discussing the general issue of why do global firms *per se* exist? This literature is predominantly driven by an understanding of manufacturing firms rather than services. This is useful in that it begins to make clear both what the main problems are with regard to the organization and coordination of global firms and how these might relate to other organizational forms. The second part of the chapter examines these debates in relation to professional services firms with a particular focus on the types of internationalization available to services companies and the factors which influence particular strategies. This section explores in particular the specific characteristics of

professional services and the way in which this impacts on internationalization. As with manufacturing, professional services can, in theory, be
delivered across national borders in a variety of ways ranging from exports
through to the establishment of affiliates in other countries. However, most
analysts recognize that services, particularly those professional services
such as management consultancy with which this chapter is concerned,
require a level of face-to-face interaction with clients that is not the case in
manufacturing. The importance in professional services of what is often
referred to as the simultaneity of production and consumption which leads
to the non-tradability (across borders) of many services therefore tends to
militate against export strategies and in principle to favour the establishment of subsidiaries or affiliates in particular localities. This is, of course,
complicated by many other factors, such as national regulations, cultural
differences and established client relationships which have in the past acted
as barriers to the establishment of subsidiary affiliates. However, in the last
two decades, an increasing number of professional service sectors (including management consultancy) have seen the emergence of 'global firms'
with strategies based on the expansion of local affiliates in multiple jurisdictions. We examine some of the issues which arise from this expansion. In
the third section, we focus more specifically on the management consultancy industry. This industry contains a number of 'global' companies with
affiliates in many different countries. What we are particularly interested in
is the way in which they work organizationally and how their structure as a
global firm is reflected in their practices. We present a broadly sceptical
account of the globalization of their practices. We argue that there is a disjuncture between the 'global' image and the 'local' reality. This in turn raises
two important areas of debate. First, why does this disjuncture exist and
how is it reproduced? Second, what does this mean for the market for management consultancy services in any particular country?

Taking a range of indicators, we argue that there are substantial organizational barriers to achieving the economies of scale, scope and learning
predicted by economic accounts of these firms. We argue that there are high
costs entailed in sustaining the global firm model in consultancy whilst the
benefits are uncertain. The reason they are sustained, therefore, is more to
do with the effect of reputational and legitimacy considerations and these
firms' abilities to convince their clients and others of their efficacy. In particular, the ambiguity and uncertainty of managerial action in general and
of the impact of management consultants on firms leads to proxy indicators of their effectiveness (Clark 1995). These proxy indicators frequently
include the characteristics of the firms themselves in relation to factors
such as size and global reach, the elite status of recruits and the influential
networks into which global consultancies are connected. Thus the global

consulting firm with high reputation and high legitimacy to other global actors (most obviously institutional investors and shareholders) becomes a preferred partner in a variety of contexts because its 'potency' rubs off on its client, whatever happens at the level of any particular consultancy project. Similarly, the client's status and reputation can work back on reinforcing that of the consultancy (for example with 'blue chip' clients). However, because of the costs of such services (related, inter alia, to the high costs of coordination in such firms) and the nature of many of them, these firms by no means sweep the board. Their services are basically directed to a particular group of large international firms[1] and large-scale projects such as outsourcing. This implies that beneath this level, there will be opportunities for many other forms of management consulting organization, not just in local settings but also across national borders. Here the specificity of national institutional contexts as an influence on the demand for management consultancy services remains especially strong. We therefore need to be cautious about analysing the management consultancy industry purely from the point of view of the large global firms. The survival and development of small firms locally, nationally and internationally modifies this picture substantially.

GLOBAL FIRMS: THE LESSONS FROM MANUFACTURING

The internationalization of manufacturing firms has been intensively studied. This reflects the economic and social context of the period from the 1950s through to the 1980s when it was the export of manufactured goods and the growth of foreign direct investment to fund production facilities overseas that characterized the increasing internationalization of the world economy. In this process, the 'multinational firm' became increasingly prominent as the organizational form which was central to internationalization. Dunning's 'eclectic theory' of the multinational explained its emergence in terms of the interaction between ownership advantages, location advantages and internalization advantages (Dunning 1993). Thus firms which had developed effective and efficient forms of production in their home context would seek to leverage their existing economies of scale and scope (ownership advantages) by expanding into new locations. Locations would be chosen according to their own advantages such as enabling the firm to avoid tariff barriers, be close to markets and consumers, reduce costs (of transport, raw materials and in some cases, labour) and enable access to new sources of capital and knowledge (location advantages). Such decisions would also be influenced by the costs and benefits of internalization of this process, as opposed to following alternative strategies such as export, licensing and franchising (internalization advantages).

During the late 1980s and 1990s, these arguments became more complex as theories of the multinational shifted away from the heavily centralized view which characterized earlier discussions towards views of the MNC as a 'heterarchy' (Hedlund 1986, 1993, 1999) or 'differentiated network' (Nohria and Ghoshal 1997). In this view, the efficiency of the MNC derived not just from its ability to leverage economies of scale and scope but also from its access to multiple sites of expertise and learning which increased its capacity for innovation. In Bartlett and Ghoshal's model (1989) of the 'transnational firm', flows of information and knowledge across different sites within the MNC via various forms of information gathering, organizational structuring and project team work gave the MNC distinctive new capacities for innovation that could not be matched inside nationally based firms. Thus in organizational terms, the problem was how to develop a structure, strategy and management system that would enable the successful achievement of this process of learning and the emergence of the 'transnational firm'. Bartlett and Ghoshal emphasize that this is not an issue of structure. The transnational cannot be reduced to a structural model along the lines of a matrix where the tensions between the three core objectives are mediated through formal reporting relationships. Responsibilities for decisions in the transnational will tend to gravitate to the most appropriate level in the organization. They state that transnationals 'decide task by task and even decision by decision where issues should be managed. Some decisions will tend to be made on a global basis, often at the corporate centre . . .; others will be the appropriate responsibility of local management. But for some issues, multiple perspectives are important and shared responsibility is necessary' (Bartlett and Ghoshal 1989, p. 209). In many ways it is the task of the headquarters to balance between different focuses and change these frequently. A hierarchical structure cannot be imposed on this diversity which is more like a 'differentiated and interdependent network . . . integrated with a flexible coordinating mechanism' (ibid., p. 210). In Hedlund's terminology, the firm moves from hierarchy to heterarchy.

This model in turn has raised three interesting debates which we will return to through the rest of this chapter (for an overview of these debates see Morgan 2005). The first which we label 'the subsidiary autonomy' debate, concerns the nature of subsidiaries and the knowledge and skills embedded in subsidiaries. In the earlier model of the MNC, subsidiaries were generally presented as 'passive'; they were recipients of the knowledge and skills transferred from the home base. In the emerging model, however, subsidiaries are perceived as 'active' with their own socially embedded skills and knowledge (Birkinshaw 1997, 2000, 2001). More recent analysis has linked this in particular to institutional arguments in which subsidiaries may have their own distinctive sets of competences and capabilities which arise from their particular local institutional context, e.g. in terms of the skill profile of their employees or the nature of the cooperation between employees and managers which exists in a particular context with consequent effects on product innovation and process improvement (Belanger et al. 1999; Kristensen and Zeitlin 2005). From this perspective, local sites are arenas of negotiation and adaptation between locally embedded actors and procedures, practices, technologies and personnel drawn from head office. Within the MNC, local sites range in the degree of autonomy and control which they construct for themselves in the internal processes of the wider firm. Some subsidiaries become active players in the construction of the firm, its boundaries, its strategies and structures whilst others are more passive.

The second debate concerns how these subsidiaries develop their distinctive capacities and capabilities and how these are coordinated across the multinational – we label this the MNC governance debate. As subsidiaries are (to varying degrees) connected into local networks of firms and institutions, it is the capacities of this local context that provides the distinctive capabilities of the subsidiary within the MNC. Solvell and Zander (1998) refer to how subsidiaries tend over time to deepen their attachment to the local context and loosen their connection to the MNC as a whole (see also Solvell 2003). Kristensen and Zeitlin (2005) identify similar aspects in their study of a particular MNC. This demonstrates that the local subsidiaries of the MNC have their own distinctive characteristics in terms of relations between managers, employees and surrounding networks of firms and local government institutions. These distinctive characteristics influence their abilities to participate in the competitive internal market. Their study raises the issue of how head office managers are able to reap the benefits of this by, on the one hand, sustaining local innovative networks and on the other hand developing mechanisms which transfer some of the skills and knowledges around the MNC itself. How is it possible to move this knowledge around effectively? How is it possible to sustain local networks whilst ensuring that there are economies of scale, e.g. from global supply management? Kristensen and Zeitlin argue that the current form of governance in most multinationals militates against this form of cooperation, instead setting sites against each other in a zero sum game. Therefore a central question for the MNC, in common with large firms generally, is how to manage and govern relations between the various constituent parts.

The third debate touches on this by bringing back in the issue of coordination costs. Can global firms do this better than markets or, more likely, inter-firm networks of various types? We label this the 'coordination costs' debate. A number of recent authors have pointed to the decline of the large integrated 'Chandlerian' firm. Langlois, for example, has argued that 'rather than seeing the continued dominance of multi-unit firms in which managerial control spans a large number of vertical stages, we are seeing a dramatic increase in vertical specialization. . . . In this respect, the visible hand – understood as managerial coordination of multiple stages of production within a corporate framework – is fading into a ghostly translucence' (Langlois 2003, p. 352; see also Lamoreaux et al. 2002; Sturgeon 2002). Langlois describes what he terms the emergence of 'a richer mix of organizational forms' in the 'new economy' arising from vertical disintegration and specialization. These forms relate to modes of coordination between different types of firms. One broad label for this whole process is, of course, Castell's 'network society' (1996) and this does indeed reflect Langlois's earlier interests in the relationship between markets, hierarchies and networks (Langlois and Robertson 1995). In the network model, firms specialize in particular positions in the value chain. So long as firms at different points can create mechanisms which allow cooperation and communication across boundaries, the network model reduces coordination and transaction costs whilst enhancing innovative capacities. More generally, higher levels of flexibility can be achieved through firms linking to others as and when needed rather than incorporating a wide range of skills inside the firm itself. The same argument can be applied to MNCs. In Kristensen and Zeitlin's analysis, the economic value added by the head office is problematic. The internal conflicts which are set in train by the competition for investment and mandates generates high amounts of effort and cost that may be economically unviable in the long run as the only way to control them is to increase standardization and this in effect undermines the *raison d'être* of the MNC as it destroys local variety and diversity, thus weakening the capacity for innovation.

These arguments point to the need to be critical and analytical about the emergence of global firms. As such firms try to access all the advantages which Dunning (1993) describes, their managerial problems increase because the maintenance of hierarchical control is difficult and counter-productive, whilst the development of heterarchical controls makes the whole system more difficult to govern in a consistent and coherent manner. It is not at all clear how these problems of governance can be resolved in ways that do not simultaneously undermine what are perceived as the main advantages of the MNC. This suggests the importance of first, being sceptical of the claims which are made for the economic efficiency of the MNC and second, being open to the emergence of different forms of coordination across national boundaries. Such a perspective has not yet been applied to the internationalization of the service sector. In the next section, we consider the dominant interpretations of internationalization amongst service firms before going on to provide our own account building on this distinctive view of multinationals.

INTERNATIONALIZATION AND THE SERVICE SECTOR

Recent decades have seen an increased interest in the internationalization of the service sector. The UNCTAD World Investment Report 2004 was entitled 'The shift towards services'. The report indicates that 60 per cent of the world's FDI stock was in this sector in 2002 (up from 25 per cent in the early 1970s). Measuring internationalization more broadly than simply through FDI in the service sector has been problematic. The UN, as part of its attempt to measure this phenomenon, has recently classified services trade into the following categories:

> Mode 1, *cross border supply* occurs when suppliers of services in one country supply services to consumers in another country without either supplier or consumer moving into the territory of the other.
> Mode 2, *consumption abroad* refers to the process by which a consumer resident in one country moves to another country to obtain a service.
> Mode 3, *commercial presence* occurs when enterprises in an economy may supply services internationally through the activities of their foreign affiliates.
> Mode 4, *presence of natural persons* describes the process by which an individual moves to the country of the consumer in order to provide a service, whether on his or her own behalf or on behalf of his or her employer. (UN 2002 Manual on Statistics on International Trade in Services quoted in Bryson et al. 2004, p. 200)

Generally the main interest in the internationalization of services refers to mode 3, the expansion of what we refer to as 'global firms' in professional services. By global firms, we mean firms that have significant affiliates or branch offices outside their home state.[2] In the professional services environment, the establishment of these global firms has occurred at different rates in different sectors. Thus by the 1990s, the firms which now make up the Big Four[3] global accountancy firms had offices in most countries in the world. In the law sector, however, globalization processes have been much slower and even now what are termed global law firms tend to have a small number of offices located in around ten key centres (or 'global cities' in Sassen's (2001) term; Taylor 2004) rather than being spread across the globe. Furthermore, it is only in the last few years that mergers of law firms across distinctive national systems have occurred as opposed to the opening up of new branch offices (Beaverstock et al. 1999, 2000; Beaverstock 2004; Morgan and Quack 2005a; Spar 1997; Warf 1997). Even then merger processes are slow and constrained occurring mostly across a small number of jurisdictions (for example the USA and the UK, the USA/UK and Germany).

Dunning (1993) identifies a range of factors within his eclectic ownership, location and internalization (OLI) advantages model that particularly influence this process in service firms. Under the category of ownership advantages, he notes the importance of quality consistency, reputation and product differentiation to keeping and winning clients in overseas location. In terms of location, the two crucial issues are whether the service is tradable and whether the regulatory framework is open. If the service is non-tradable across borders and requires co-presence, then the service firm has to locate in order to win business, even where it has an existing long-standing relationship with the client but in different national contexts. However, this requires that there are no regulatory barriers to foreign presence, which was not the case in many professional services until recently. Finally where the service requires customization and interaction with the client, it cannot really be delivered by franchise or licensing. The firm has to retain the skills in-house and to deliver them to clients itself. Under these circumstances, Dunning predicts the emergence of international professional service firms.

Other authors have linked ownership and internalization advantages to the issues of knowledge and learning in two ways; first, that of leverage and second, that of innovation. Leverage implies that resources established in one context can be reused in other contexts. The degree to which professional services fit this is variable depending on the types of business. For example, at first sight, a law firm in London has no obvious resources which would enable it to compete effectively in Shanghai. However, things are rarely as simple as much commercial business between multinationals in Shanghai and the law associated with it is likely to be based on English law. Therefore the more accurate analysis would be that the UK law firm has no advantages to bring to a situation where it is competing for local clients but it has many where it is working for UK MNCs or other firms that are having to contract under English law. Furthermore, once established it may be able to compete for local business on the grounds of its international reputation (assuming it has employed local lawyers in its office).

The issue of reputation is crucial in this context. As Teece states 'while reputational capital is certainly not unique to professional service firms, it is frequently the most important asset. This is because other methods of selling – for instance, advertising – are usually quite ineffective' (Teece 2003, p. 902). Aharoni states that:

> the ability to give services globally is part and parcel of a perception of high reputation and an indicator of competence and commitment to service the

customer. Global firms are able to transfer reputation from one geographical market to another. In fact, the major advantage of the giant global accounting, consulting or advertising firm is that clients believe these firms connote high quality. (Aharoni 2000, pp. 127–8)

Professional service firms rely especially on word of mouth recommendations or 'networked trust' (Gluckler and Armbruster 2003) where reputation is gained and reinforced by connections to and endorsements from other powerful actors within particular corporate networks, e.g. around Wall Street and the City of London for law firms and around 'blue-chip' corporate clients for management consultancies. Therefore professional service firms may see internationalization as an opportunity to leverage their existing resources further by building and developing their existing reputation.

The first way in which firms achieve this is often by following their clients abroad so that their initial contacts and networks can be leveraged further in new markets. This activity has been labelled as 'client following' and it may be accompanied by or lead onto 'market-seeking' strategies, that is, looking for new clients in overseas markets. An alternative or complementary approach is to recruit or 'poach' senior consultants from other firms with existing client contacts in a particular region (see Pinault 2001).

A rather more complex but equally important element concerns innovation. It is generally argued that innovation in professional service firms is an outcome of interaction with clients (rather than from the establishment of the equivalent of R&D labs or the linkage with outside 'incubator' firms that characterizes some areas such as biotechnology and IT) (see for example Mills and Morris 1986; Lowendahl et al. 2001). Clients' problems generally have elements of novelty as well as routine. Professional service firms therefore tend to balance off the advantages of standardization with those of the customization of solutions depending on the sector of professional services in which they are working, their strategic positioning and their structural configuration. Extreme or, rather, explicit standardization runs the risk first, of a loss to reputation as clients feel 'cheated' and second, that opportunities to identify new products and processes which can be sold on to other clients are missed. Professional service firms are likely, therefore, to look for ways in which new opportunities can be identified. Internationalization in particular has the potential, as Bartlett and Ghoshal (1989) described in relation to manufacturing firms, to open up new innovative ways of doing things by diversifying the nature of the clients and the problems with which the firm deals. As Lowendahl states: 'in PSFs [professional service firms] the competitive advantage, if achieved, results from the ability of the firm to continuously tap into the knowledge developed in all relevant centres of the world, regardless of the local market potential in these knowledge centres' (Lowendahl 2000, p. 152).

These firm-level drivers towards internationalization have gradually got stronger as the broader environment has changed. Key elements here have been the emergence of forms of international standardization embedded in discourses of globalization. These revolve around arguments about the increasing integration of financial markets, the development of common global standards of business knowledge and techniques propounded in the media, business schools and management consultancies, and the emergence of formal and informal modes of international regulation (Morgan 2001a, 2001b; Thrift 1998). These discourses of globalization exert a powerful influence on many actors, providing for them both an understanding of the contemporary social and economic environment and also a way of acting in that environment. Large global professional services firms and their clients are key participants both in constructing this discourse and enacting it as a social reality in the face of potential national resistance (for example through spreading particular conceptions of law, Dezelay and Garth 1996, 2002a, 2002b; international standards of accounting, Arnold 2004; Botzem and Quack, forthcoming, and particular forms of management knowledge, Sahlin-Andersson and Engwall 2003). The global presence of professional service firms is legitimated by the discourse of open borders and common practices whilst the commonality of practices and discourses is reinforced by their actions for clients, both in the private sector and inside particular states or cross-national institutions such as the IMF, the World Bank and the WTO.

In conclusion, there are a number of drivers towards the internationalization of professional service firms. These can be partially explained in relation to previous models of internationalization but for a variety of reasons, a number of other factors come into play. The issue of co-presence and the centrality of client–professional interaction is particularly important. This affects how firms internationalize in order to leverage their existing resources and build new products and innovations whilst maintaining and developing their global reputation. In the following section, we look at these issues particularly from the point of view of the management consultancy industry.

THE MANAGEMENT CONSULTANCY INDUSTRY AND INTERNATIONALIZATION

In this section, we first, concentrate on global management consultancy firms and how they organize themselves as 'transnationals'. In this respect our concern is with understanding the way in which they represent themselves as global and how this is seen to relate to their capabilities to serve clients. Second, we look at this more critically and consider more specifically what 'being global' means in this context. We relate this to the broader problems of multinational organization described earlier as 'the subsidiary autonomy' debate, the 'governance' debate and the 'coordination costs' debate.

In organizational terms, global management consultancies vary in their strategy and structure (Kipping 2002). Two broad categories can be identified though there is some overlap between them. What Kipping refers to as the second-wave consultancy groups (following the first wave originating out of the scientific management movement) are the prestigious 'strategy' consulting groups such as McKinsey and Boston Consulting Group. These firms became international in the 1950s and 1960s spreading their offices particularly into Europe as part of the broader expansion of US management ideas into the European environment that occurred at this time (Djelic 1998; McKenna et al. 2003). These firms were built on the partnership form of governance and have continued to retain this. Although their business interests have spread from pure 'strategy' research, they tend to remain distinctive in culture and approach from the other main grouping which emerged out of the big accounting firms in the 1980s (for an example of these differences in action see Empson 2001).

By size and geographical spread, the largest and most global management consultancies are those which originated inside accounting firms. In Kipping's terminology these are the third wave of consultancy business. The consultancy business of the accounting firms grew particularly on the basis of selling advice in the light of knowledge gained about management problems through the auditing process. This tended to be concerned with more operational issues than that of the strategy consultants, for example in terms of business processes and IT integration. More lately, the shift towards outsourcing of various functions, particularly associated with IT and various forms of payment, invoice and accounting procedures has added a third major leg to their business.

In organizational terms, the establishment of these businesses increasingly created tensions within the accounting firms. One aspect related to the relations between qualified accountants and the management consultants, issues which revolved around reward packages, prestige and status and generally power within the firm. Tension also emerged partly within the firm but also in the broader regulatory context concerning the impact of the drive for consulting income on the probity of auditing activities. In particular, it was increasingly claimed firms were being less objective about audits than was expected because they were keen to sell on consultancy services to their audit clients and did not wish to lose the potential for this more profitable business by being overly critical of the company on these issues (Stevens 1981). The result has been the splitting off of consultancies from the accounting firms. The earliest example of this was when Accenture split off from the now deceased Arthur Andersen accounting company. More recently and more closely related to impact of regulatory scandals such as Enron, there has been increased pressure on the accounting firms to sell off their consulting arms. Cap Gemini bought Ernst and Young's consulting arm in 1999 and in 2002 PwC Consulting was bought by IBM. KPMG's accounting arm was given a NASDAQ IPO in 2001 and following the acquisition of a number of ex-Andersen businesses changed its name to BearingPoint Inc in 2002. Interestingly, the separation of these units from their accounting firms meant a shift in governance, from the partnership structure to a shareholder structure and quotation on stock markets. Thus the two main areas of global consulting are now differentiated more strongly by their governance structure than by their broader strategic purposes.

What are the advantages which these global management consultancies bring to their tasks? Clearly this partly depends on the nature of the task, for example, whether it predominantly requires work in one location or whether it spreads across national boundaries. Either way, however, the global management consultancy will claim advantages over local or national firms purely by virtue of its 'global reach' and the processes which are necessary to achieve this. As Backlund and Werr in their study of the websites of global consulting firms state:

> All firms underline that they are a global or international consulting firm. The global character of these companies is also reflected in their recurring descriptions of being 'world leaders' in their respective businesses. . . . More or less implicit in these references to globality is the view that a global reach contributes to the consultancies' knowledge stock. (Backlund and Werr 2004 pp. 31–2)

The companies construct their global identity as key to their ability to serve their clients in a number of ways. First, in addition to general size advantages such as having potentially large numbers of staff available to work on large projects and as a back up if resource requirements change, there is the issue of experience and the global management consultancy firm's claim to have the experience of managing complex cross-national projects. It will claim benefit from what has been termed the 'economics of repetition', i.e. the existence of tried and tested processes inside the firm which enable it to pull together cross-national teams as a matter of routine. Second, the global firm will claim benefit from the stored expertise and knowledge of its consultants across the world who may have faced similar problems. Such firms have various techniques for capturing this knowledge ranging from 'knowledge management databases' which store cases and examples through to the identification of particular individuals as experts or centres of excellence (see below and Roberts 2004) in the problem area, part of whose job is to resource the organization more generally with this knowledge as and when required (Rudolph and Okech 2003).

Overall, the global management consultancy can claim to be advantaged by economies of scale (it has a greater source of specialized knowledge than comparable national firms), economies of scope (it has experience of a greater range of connected problems and solutions that can be leveraged by consultants to extend and develop the scope of activities), economies of repetition (it has an established set of routines that can be set into motion as and when projects are first mooted) and economies of learning (it has routines for capturing, managing and storing knowledge that can be used again). On top of this the global management consultancy has the crucial advantage of reputation. In an area like consulting where there is no standardized knowledge base, reputation can be built through association and global consulting firms tend to be highly involved in this on a number of levels. One level concerns the process of recruitment itself. Global consultancy firms target the top business schools in the USA for their junior recruits or subsequently send them to such schools. Armbruster, for example, states:

> The highly selective recruitment of outstanding individuals provided the consulting industry with a considerable touch of intellectual elitism. . . . (and) is still considered a way of propagating the notion of an intellectual elite to the business environment . . . [Because] the quality of consulting services is difficult to determine . . . management consultancies need to signal the quality of their services by substitutive means, among which is a highly selective hiring process. (Armbruster 2004, p. 1259)

Global management consultancies are keen to have long lists of applicants to enhance the elitism of those who are selected. Another way in which this is reinforced is through the almost exclusive reliance of the leading management consultancy firms on 'the case study interview – a job interview in which an abbreviated form of a Harvard Business School case study is posed to the candidate' (Armbruster 2004, p. 1250).

> Being a tool that is only used in management consulting, the case study serves as a signifier of otherness and analytical skills. Also . . . the selection procedure is associated with Harvard Business School and thus symbolizes special business training and competence. (Armbruster 2004, p. 1260)

The fact that in most cases the selectors are themselves not trained HR practitioners but rather working consultants in the firm who 'have only received training in one-day or half-day courses' (op cit., p. 1258) reinforces the sense of an elite self-contained group. Alvesson and Karreman argue that this sort of recruitment policy and procedure contributes to 'producing a fairly homogeneous work force' (2004b, p. 431). In addition, strong cultural control in terms of selection, induction and training in corporate methods of analysis reinforces this effect (Roberts 2004). This is even evident in cases where firms recruit from outside the firm, at more senior levels for example, where cultural fit is seen as important (Pinault 2001).

The management of reputation through recruitment and selection can be reinforced by formal links with business schools, such as supporting research or endowing chairs or sponsoring their staff onto MBA degrees. Management consultancies can also act 'like' universities and academics, for example, by creating their own in-house learning institutes which draw in academics from universities or by encouraging their employers to present conference papers or publish in academic journals. They can also create their own publicly available journal along academic lines, for example, *McKinsey Quarterly*. Consultancies can also draw reputation from the secondments which they support, for example, into national or international governmental agencies as well as from the jobs which their consultants go on to after they leave the firm. In cases such as McKinsey, there are active efforts to sustain the McKinsey networks and publicize the connections that the firm has to important people and institutions. Reputation is also built through having the right sorts of firms as clients and having access to the right sorts of people within such firms and even recruiting from them or other consultancies with such client links.

Given all these advantages, how do these processes work out in practice? We look at this mainly from the point of view of how aspirations to share knowledge relate to other organizational processes in the firm. A useful case study which illuminates some key issues has been produced by Fenton and Pettigrew (2000) in their analysis of how Coopers and Lybrand (Europe)[4] sought to develop their expertise in the area of pharmaceuticals by creating a cross-European network of expertise consisting of around 200 consultants. They state that 'in 1996 the pharma network was formally established as a business development group. It was required to develop an integrated business plan across Europe'. In practice, however, a number of substantial barriers emerged to this network becoming an effective force within the firm. The first and most significant point was that the main power base in the firm was at the national level. This was strongly associated with the fact that the firm was in effect a federation of national partnerships. Individuals were members of particular national partnerships and it was at this level that careers and rewards were organized. Partners themselves, for example, were rewarded on the basis of the earnings which their national partnership achieved. Consulting projects were constructed at the national level even where they were with multinational firms. In all client engagements, there would be a lead partner and it would be the lead partner who would determine the membership of the team working on the project and how the division of labour was to be constructed. The purpose of the pharma network was to ensure that teams in this area were built from the best consultants no matter what their location. However, according to Fenton and Pettigrew, 'there was a strong perception that national firm requirements often took precedence over opportunities in the marketplace . . . the feeling was that they were still operating as a national organization rather than a European network' (Fenton and Pettigrew 2000, p. 100). Much of this was related to the fact that the ultimate profit centres and power centres in the firm were the national partnerships and within this, there were certain partnerships which were central and others which felt peripheral. For individuals it was most important to contribute to the profitability of the national firm and in this way earn high evaluations and promotion. For some firms, the pharma network was peripheral and their consultants were unlikely to become involved. For other firms, where senior partners were enthusiastic backers of the pharma network, consultants needed to be involved. In all these cases, the degree of knowledge sharing was contingent on other factors more deeply embedded at the national (that is, organizational structural) level in terms of careers and rewards.

This is reinforced by deeper studies of how consultants are monitored and controlled. Alvesson and Karreman, for example, refer to the 'technocratic' system of control in their management consulting case study (Alvesson and Karreman 2004a, 2004b). A number of their points reinforce our argument about the importance of local/national contexts. First, they state that selection on to projects requires that juniors 'have good relationships with what they perceive as good project managers so that the latter will choose/offer them for projects that are attractive to work in' (Alvesson and Karreman 2004b, p. 432). Second, employees are evaluated on a frequent basis by their nearest boss, usually their project manager who grades them on performance. Third, they are provided with a mentor, usually a partner, who offers advice on career advancement. All of these processes are dominated by the local office, thus cutting across attempts to achieve cooperation across offices in different countries or regions.[5]

Much of this is associated with the intertwined issues of careers and project budgeting in management consultancies. The evaluation and monitoring is part of a broader process of ensuring that new members of the firm see that their elite status on recruitment has to be sustained by high levels of commitment to the firm and effectiveness in fitting in with and contributing to the success of projects, especially in terms of generating repeat business or 'sell on'. The large consultancies therefore famously operate what are known as up or out promotion systems. In Nanda and Morrell's study of BCG, they describe the expected timescale to move from associate to senior Vice-President (the BCG partner level) as ten years. Thus the anticipated period on each of the intermediary rungs of the ladder (consultant, project leader, manager) is around two to three years. An employee failing to be promoted in that time is basically counselled to leave the firm and given help in moving to another job, not least as they may become future clients. The outcome of this is that roughly one in ten of the new entrants is likely to reach partner level whilst the others will leave. For the individual this creates enormous pressures to perform, reflected in working long hours, a willingness to travel away from home and a general high level of commitment to the firm (see Alvesson 2000 for an analysis of the tensions involved in this process; also Alvesson 2001). It also creates a strong pressure to focus on relationships with particular key individuals who are likely to be based in one location, the local office. As one consultant told Nanda and Morrell in discussing whether to apply for a prestigious overseas relocation position (known as the Ambassador programme in BCG), 'by relocating for some time, I might be cutting myself out of the loop. How will I get back into a favourable position in the staffing pool? Will my next promotion be delayed?'(Nanda and Morrell 2004, p. 10). This reflects a broader problem in professional service firms concerning issues of relocation. In a context where relationships are crucial – with partners, with clients, with managers and consultants – movement out of a known environment to an office in another national partnership cuts the individual off from these sources of support and makes them vulnerable. For this reason, although these firms are characterized by high levels of mobility across national borders for purposes of doing projects with multinational firms, there is a general reluctance to make a more long-term move. This may be a possibility early in the career or as a carrot for a rather quicker promotion, especially if skills match those required by a prestige client, but at partner level it is less likely to be undertaken.

The argument here is not that the formal procedures of the firm discourage working on cross-national projects and sharing knowledge. In terms of the appraisals and evaluations which employees undergo, it seems clear that emphasis is given to these aspects of performance although this varies in practice. The issue is more to do with how the broader environment of career affects the individual in terms of (a) the willingness to share (where career advancement is in effect a tournament between peers) and (b) the amount of time which is available above and beyond project time to make this effective, not least given the difficulties of translating tacit to explicit knowledge. If lower-level employees are involved in a tournament, what are the advantages of sharing knowledge particularly where time pressures are so tight? Of course, this may be built into an appraisal system but then it becomes difficult to disentangle the effects of the measures of sharing from the reality of sharing. Thus if sharing is measured, amongst other things, by the number of postings one makes to the knowledge management system then the tendency is to post more notices to the system whether or not they are useful. Such opportunistic behaviour can in theory be corrected by a further level of monitoring – how often is a posting used by others in the firm and how do they evaluate it? But all of this introduces new levels of complexity, bureaucracy and costs in order to correct what is a central part of the system, that is, the tournament process and the effect it has on commitment and hours of work and thus on budgets, margins and profitability.

This relates to the second point which is that of project budgeting and accounting for time. Alvesson and Karreman (2004b) describe the role of partners primarily in terms of selling projects and developing the client agreement. Their rewards (that is, how many units – shares – they are allowed to purchase in the following year) are determined by the volumes of business sold and the margin achieved on that business. The issue of maximizing the margin and monitoring the budget is the responsibility of the project manager and this is done through planning the project in terms of manpower, resources and cost. This creates a strong framework within which the consultants' billable hours are strictly accounted for, leaving very little slack. On the contrary, as Alvesson and Karreman illustrate, billable hours often underestimate the actual hours undertaken by consultants, an issue that exercises the consultants greatly but is swept under the carpet by the project manager for the obvious reason that if all hours were accounted for, the price of the job would have to increase and the client might go elsewhere (Alvesson and Karreman 2004b, pp. 433–6). In effect, partners rely on their ability to extract more hours from their juniors than they formally request; juniors conform, if reluctantly, because they know their reputation as a good team player (and therefore their promotion) is dependent on this.

Dunford links some of these issues to the effectiveness of formal knowledge management systems for sharing experiences. He states that the quality of information in such systems 'may be impaired at a very basic stage by consultants failing to feed information into their firm's system, This may be due to something as fundamental as time pressure' (Dunford 2000, p. 297). He also points to the way in which performance evaluations primarily based on billable hours 'discouraged consultants from taking time to package and share new insights' (ibid., p. 298). His overall conclusion is that 'a lot of the investment in knowledge management is an act of faith' (ibid., p. 301).

This relates to a broader point about the nature of the global management firms. Their basic organizational unit is the national partnership and the local offices within that framework. This is where rewards and careers at all levels are ultimately determined. It is where business is generated and projects managed. Where national partnerships gain most clearly from their membership of a global firm is in reputational terms. In practical terms of sharing information, knowledge and practice across boundaries the gains are more limited although there is some benefit from drawing on (functional/client) specialists, albeit at a distance, by phone, e-mail and video for example. Clearly many consultancies share a standard model of work practices and they also share a general registry of expertise and knowledge. However, to make this more than a limited benefit requires efforts at integration, cooperation and joint learning that have one obvious drawback as far as all members of the firm are concerned. That drawback is that they generate high overheads. For example, establishing an in-house university or an international management development is very expensive and involves taking consultants off billable activity. Establishing a global network of expertise that can be called upon at any point to advise on projects is expensive if it extends beyond a list of names and telephone numbers. Capturing more of the firm's revenue to support activities which are not directly revenue-creating goes against the grain of these highly focused, budget-driven organizations as well as against the interests of most of the individuals inside them. Thus firms can espouse the rhetoric of global knowledge management systems and global integration of teams and knowledge, but in reality they find it very difficult to deliver this except insofar as their services are standardized. The firms with the highest margins on business (that is the strategy firms) are likely to be better able to move in this direction than the mass consultancy firms which tend to focus on lower margin projects. However, contrary to their projected image, even the strategy houses offer standardized forms of analysis so whatever the type of the firm, the problems are large. As Fenton and Pettigrew state 'our probing within the Professional service organizations revealed a discernible system-wide imbalance in the degree of integration within their networks because of a variety of confounding factors' (Fenton and Pettigrew 2003, p. 230). These confounding factors include contextual issues (differences across national markets, skill differences between partners), historical factors (concerned with how different national groups saw the historical inequalities between partnerships) and social factors. On social factors they state:

> None of the PSOs [professional services organizations] had developed effective sanctions beyond peer group pressure; similarly there was little in the way of

reward and recognition systems in place to encourage positive behaviour in the changed firms. (Fenton and Pettigrew 2003, p. 232)

In addition, it is important to note the role of clients in shaping consulting practices, especially the more sophisticated, innovative or active ones who may well resist the offer of a standard consulting approach (Sturdy 1997).

In conclusion, then, our argument is that just as with global multinational firms in the manufacturing area there are good reasons for believing that there are inherent limitations to the growth of global firms in management consultancy. To return to our three debates in MNCs on subsidiary auton-omy, governance of the global entity and coordination costs, we conclude as follows. First, 'subsidiaries' tend to have high levels of autonomy. Indeed, up until recently the global management consultancies were run as feder-ations of national partnerships.[6] It is at the national level that business is won and the main rewards are distributed. 'Global' controls on the national context vary in terms of their intensity and effectiveness. Standard method-ologies in particular areas and shared databases of expertise may be one useful element but most crucial at the global level is the reputational element. Local offices need to sustain this reputation and although this creates some constraints, there are many areas of freedom that remain. Thus we conclude that global management consultancies have relatively high levels of subsidiary autonomy, for example, to develop new business areas, new techniques etc.[7] In terms of governance, our general conclusion is that 'global governance' in such firms is weak and as a result there may fre-quently arise tensions between different national practices in terms of how far 'peripheral' practices become involved in the large projects won in the big consulting markets. The issue of how far work is contracted out from a lead partner in one country to consultants in other countries is unclear. Barrett et al.'s (2005) analysis of how international audits are coordinated across different national practices within the same global firm reveals that this can involve complex political and economic bargaining between more or less powerful national partnerships (for similar arguments see also Rose and Hinings 1999 which looks at global business advisory firms which they label GBAFs, that is accounting firms with consultancy arms). Overall, therefore, the potential for political action within the global structure of the firm seems high, though this may be mitigated by two factors. First, in most of the global consultancies, the US partnership is dominant by virtue of the strength of the US consulting markets and the central role of the US part-nership in sustaining reputation and cultural control. Second, the relatively limited resources placed at the global level means that there is not so much worth fighting for as there is in manufacturing firms where the global head-quarters effectively controls the surplus as well as access to investment funds through the capital markets.[8] Finally the issue of coordination costs across national boundaries impacts on two levels. First, it is a considerable problem at the level of projects and creating integrated learning. The tight budgetary control exercised on activities through the billable hours system militates against expenditure on cross-national coordination. Second, the structure of national partnership control means that the global level is relatively weak and dependent for resources. It can therefore only achieve limited objectives in terms of 'global activities'. Thus coordination costs are kept relatively low in these systems but the result is that the global integration which is proclaimed as an essential part of their competitive advantage is not actually achieved.

DISCUSSION AND CONCLUSION

Our analysis of the global management consultancy firm suggests that there are organizational problems which mean that it is does not deliver on the 'transnational solution' as described by Bartlett and Ghoshal. Where it delivers is in terms of reputation and as Aharoni (2000) and others have pointed out this is essential not least because in the arena of professional services, there are high levels of uncertainty and ambiguity concerning the effectiveness of consultancy interventions (Sturdy 1997; also Clark and Fincham 2002). Reputation works for both clients and consultants. For clients it gives them some security in their decisions and the effectiveness of those decisions even where this is difficult to measure. For consultants, it imbues them with a sense of being members of an elite with access to privileged and specialist knowledge and confidence in its application. Managing reputation on a global scale works for national partnerships as well, enabling them relative autonomy so long as they stick to the essential elements that make up the reputation, i.e. the recruitment procedures, the up or out promotion system, the maintenance of prestigious client lists and so on.

On the other hand, if the outcome of this is that the nature of what is delivered is not distinctively 'transnational', it opens up the question of whether other forms of management consultancy organization might not still be able to grow and prosper in particular areas of the market. Kipping et al. reinforce this point in that they note that 'the consultancy fields in most European countries continue to be dominated by individual or very small service providers of domestic origins' (Kipping et al. 2003, p. 37; see Rudolph and Okech 2004 for the German case). The nature of national management consultancy markets is clearly different and continues to affect both the sort of consultancy services required and more particularly the nature of the clients, the firms and public organizations requiring those services (see the interesting collection of papers in Kipping and Engwall 2002 which look at different national contexts). Lowendahl, for example, distinguishes between global clients, local clients with 'global problems' and local clients with local 'problems' (Lowendahl 2000, p. 153–7). Whilst one would obviously expect a dominance of global management consult-ancies with global clients and the dominance of local consultancies over local clients, there is an area in the middle where there may be a great deal of competition. This is, in part, an issue of size in that global firms are also typically large firms and can reduce the risk to clients by having a relatively flexible resource base in terms of the number and skills of available con-sultants, but there are also cost implications. Global management consult-ancy firms incur large costs in their recruitment and career management. They are unlikely to be able to compete on cost terms with more local or national firms so clients will have to balance out these cost considerations with reputational and competence considerations before deciding how to allocate business. This is also, in part, an issue of the relationship to clients and how well being part of a global management consultancy may facili-tate that. For example, one of the ways in which 'national' firms emerge is through consultants moving out of the global business to set up their own firm. Their international experience is clearly beneficial in this but their independence from the global firm and their potential network derived both from the global firm and their broader social, political and educa-tional contacts may be relevant to them winning clients against the global firm. This independence often enabled them to adapt the dominant US message and the US techniques of the global firm even more closely to the local circumstances. Local high-status employees could use their own high status and that of the firm which they had just left to make their way into local networks. There is evidence that one issue for these types of firms is that they begin to imitate too strongly their US origins leading them into expansion plans which eventually end up with them seeking out a merger with one of the global firms or being sought out on the basis of (re-)cap-turing their client base. Such high-profile national firms have therefore tended to decline in significance over the last decade particularly as they have found that their home-based MNC clients are more interested in inter-national advice provided by high status international management consul-tancies with greater resources. Therefore the space for this sort of firm has been severely squeezed.

On the other hand there is clearly the continued existence of a sub-strata of local firms with smaller scale but higher specialization that can seek to service national clients. It is this grouping which is least researched. There is a tendency to assume that this grouping basically serves small and medium-sized firms rather than multinationals and therefore is not really of interest to debates on internationalization. However, in our view, this case is not proven. There are some indications in management consultancy (as there are in law) that highly specialist, small-scale 'boutique' operations can gain business in large multinationals in particular areas. Such consultancies can legitimate the claim to customization and 'unique solutions' in a way which the global firms cannot so easily do. Whilst they lack the flexibility derived from scale advantages, they can keep overheads relatively low and ensure that their seniors work more directly with the clients throughout the process. What is less clear is whether such firms can at least in part leverage their own international capabilities by entering into networks. Historically, a lot of cross-border trade[9] in professional services has been conducted through networks of referrals and limited movements of high-reputation individuals. In this model, clients of professional services firm A in country Z requires advice about its activities in country Y. Firm A is in a network with other independent firms including firm B in country Y. The client's problem is referred to firm B by firm A. In many networks, there will be no fee for the referral as the network works on the basis of long-term reciprocity so that firm A will expect in the future to receive a referral from firm B. Firm B bills the client directly. Such networks can take many forms in terms of formality, virtuality, permanence, mechanisms of control and governance. However, they are unlikely to appear in FDI statistics (because there is no flow of capital, just information) even though they may well be the most common way in which small and medium-sized firms are able to seek advice on international issues. The idea of referral networks in management consultancy is relatively unexplored. Clearly the management of such networks would need to be very careful as alliances with the 'wrong sort' of partners could destroy credibility. On the other hand in the right circumstances it might be a possible alternative to the global firm though there is little research on this issue yet which can provide detailed information on which to make a judgement. The network may provide a way for smaller more specialist firms to serve international clients as well as smaller firms with international requirements. Another aspect of this relates to the 'virtual firm'. There are a number of examples of nationally based consultancy operations which are, in effect, 'virtual firms'. By this we mean that the firm itself is simply an administrative centre; it does not have its own complement of consultants but rather has a list of other firms (often very small operations down to one-person businesses) on whom it can call when an enquiry is made. In effect it acts as a dating agency bringing together the client and the consultant. However, the client may be unaware that this is what happens and may believe that the consultant is an employee of the firm with which the initial contact was made. The 'virtual firm' model may appear highly precarious as it is unclear how standards of service are monitored and protected although the professional consulting associations are more evident at the level of small- and medium-sized firms. On the other hand it reduces overhead costs significantly and has the potential for reaching beyond the national context.

In conclusion, our argument is that the internationalization of management consultancy firms should not be studied solely from the perspective of the global firms. If we look in more detail at these firms we can see that the benefits which they get from internationalization are primarily derived from reputation. Underneath this, the global firms remain federations of national partnerships with limited integration or knowledge sharing. In particular national markets, national firms are likely to be able to link more carefully to the distinctive characteristics of those contexts, most obviously in terms of their relations with small and medium-sized enterprises but also in other ways through their ability to adapt broader messages and techniques to local contexts. Furthermore, national firms are not necessarily confined within national markets. The possibility of network connections or of the creation of virtual firms linking independent consultants to a common marketing arm also exists. It will be an interesting project for future research to examine more carefully the extent and significance of such alternative international forms.

NOTES

1. We should also note here the increasingly important role these global consultancies play in advising governments, transnational bodies and large-scale voluntary international organizations on a range of issues.
2. In this context, we are not so concerned with the degree of globalization per se or with whether these firms are best described as 'regional multinationals' rather than global firms (see Morgan and Quack 2005a for a more nuanced discussion of this in relation to law firms).
3. By the time of the collapse of Andersen, the big accountancy firms were four in number reducing from the Big Six in the early 1990s and the Big Five in the late 1990s.
4. Their study is based in the mid-1990s before Price Waterhouse and Coopers and Lybrand merged to form PWC and the Big Six became the Big Five.
5. Similar processes are described in the Harvard case study by Nanda and Morrell on consultants in Boston Consulting Group (Nanda and Morrell 2004).
6. It will be interesting to see how the shift to plc status for those companies spun off from the accounting firms may impact on these issues in the future.
7. For similar arguments developed from a rather different perspective see Jones 2002, 2003.
8. Rose and Hinings (1999) provide an interesting discussion of the international level of governance in their analysis of GBAFs.
9. Such referral networks are also relevant *within* societies, particularly where regulations have constrained a firm from establishing multiple offices across different localities (as has been the case for law firms in many countries) but also in conditions of geographical distance.

REFERENCES

Aharoni, Y. (2000), 'The role of reputation in global professional business services', in Y. Aharoni and L. Nachum (eds), *Globalization of Services*, London: Routledge, pp. 125–41.

Aharoni, Y. and L. Nachum (eds) (2000), *Globalization of Services*, London: Routledge.

Alvesson, M. (2000), 'Social identity and the problem of loyalty in knowledge-intensive companies', *Journal of Management Studies*, **37** (8), 1101–23.

Alvesson, M. (2001), 'Knowledge work: ambiguity, image and identity', *Human Relations*, **54** (7), 863–86.

Alvesson, M. and D. Karreman (2004a), 'Cages in tandem: management control, social identity and identification in a knowledge-intensive firm', *Organization*, **11** (1), 149–75.

Alvesson, M. and D. Karreman (2004b), 'Interfaces of control: technocratic and socio-ideological control in a global management consultancy firm', *Accounting, Organizations and Society*, **29** (3–4), 423–44.

Armbruster, T. (2004), 'Rationality and its symbols: signalling effects and subjectification in management consulting', *Journal of Management Studies*, **41** (8), 1247–69.

Arnold, P. (2005), 'Disciplining domestic regulation: the World Trade Organization and the market for professional services', *Accounting, Organizations and Society*, **30** (4), 299–330.

Backlund, J. and A. Werr (2004), 'The construction of global management consulting – a study of consultancies' web presentations', in A.F. Buono (ed.), *Creative Consulting: Innovative Perspectives on Management Consulting*, Greenwich, CT: Information Age Publishing.

Barrett, M., D. Cooper and K. Jamal (2005), 'Globalization and the coordinating of work in multinational audits', *Accounting, Organizations and Society*, **30** (1), 1–24.

Bartlett, C.A. and S. Ghoshal (1989), *Managing across Borders: The Transnational Solution*, London: Century Business.

Beaverstock, J. (2004), ' "Managing across borders": knowledge management and expatriation in professional service legal firms', *Journal of Economic Geography*, **4** (2), 157–79.

Beaverstock, J., R.G. Smith and P. Taylor (1999), 'The long arm of the law: London's law firms in a globalizing world economy', *Environment and Planning A*, **31** (10), 1857–76.

Beaverstock, J., R.G. Smith and P. Taylor (2000), 'Geographies of globalization: US law firms in world cities', *Urban Geography*, **21** (2), 95–120.

Belanger, J., C. Berggren, T. Bjorkman and C. Kohler (1999), *Being Local and Worldwide: ABB and the Challenge of Global Management*, Ithaca, NY: ILR imprint, Cornell University Press.

Birkinshaw, J. (1997), 'Entrepreneurship in multinational corporations: the characteristics of subsidiary initiatives', *Strategic Management Journal*, **18** (3), 207–29.

Birkinshaw, J. (2000), *Entrepreneurship in the Global Firm*, London: Sage.

Birkinshaw, J. (2001), 'Strategy and management in MNE subsidiaries', in A. Rugman and T.L. Brewer (eds), *The Oxford Handbook of International Business*, Oxford: Oxford University Press.

Botzem, S. and S. Quack (forthcoming), 'Contested rules and shifting boundaries: international standard setting in accounting', in M.-L. Djelic and K. Sahlin-Andersson (eds), *Transnational Regulations in the Making*, Cambridge: Cambridge University Press.

Brock, D., M. Powell and C.R. Hinings (eds) (1999), *Restructuring the Professional Organization*, London: Routledge.

Bryson, J.R., P.W. Daniels and B. Warf (2004), *Service Worlds: People, Organisations, Technologies*, London: Routledge.

Castells, M. (1996), *The Rise of the Network Society*, Oxford: Blackwell.

Clark, T. (1995), *Managing Consultants*, Buckingham: Open University Press.

Clark, T. and R. Fincham (2002), *Critical Consulting: New Perspectives on the Management Advice Industry*, Oxford: Blackwell.

Dezelay, Y. and B. Garth (1996), *Dealing in Virtue*, Chicago: University of Chicago Press.

Dezelay, Y. and B. Garth (2002a), *The Internationalization of Palace Wars*, Chicago: University of Chicago Press.

Dezelay, Y. and B. Garth (eds) (2002b), *Global Prescriptions*, Ann Arbor, MI: University of Michigan Press.

Djelic, M.-L. (1998), *Exporting the American Model*, Oxford: Oxford University Press.

Dunford, R. (2000), 'Key challenges in the search for the effective management of knowledge in management consulting firms', *Journal of Knowledge Management*, **4** (4), 295–302.

Dunning, J. (1993), 'The globalization of service activities', in J. Dunning, *Globalization of Business*, London: Routledge, pp. 242–84.

Empson, L. (2001), 'Fear of exploitation and fear of contamination: impediments to knowledge transfer in mergers between professional service firms', *Human Relations*, **54** (7), 839–62.

Fenton, E.M. and A.M. Pettigrew (2000), 'The role of social mechanisms in an emerging network: the case of the pharmaceutical network in Coopers and Lybrand Europe', in A.M. Pettigrew and E.M. Fenton (eds), *The Innovating Organization*, London: Sage, pp. 82–116.

Fenton, E.M. and A.M. Pettigrew (2003), 'Complementary change: towards global integration in four professional service organizations', in A.M. Pettigrew, R. Whittington, L. Melin, C. Sanchez-Runde, F. van den Bosch, W. Ruoigok and T. Numagami, (eds), *Innovative Forms of Organizing,* London: Sage, pp. 208–39.

Geppert, M., D. Matten and K. Williams (2003), *Challenges for European Management in a Global Context*, London: Palgrave Macmillan.

Giddens, A. (1990), *The Consequences of Modernity*, Cambridge: Polity Press.

Gluckler, J. and T. Armbruster (2003), 'Bridging uncertainty in management consulting – the mechanisms of trust and networked reputation', *Organization Studies*, **24** (2), 269–97.

Hedlund, G. (1986), 'The hypermodern MNC – a heterarchy?', *Human Resource Management*, **25** (1), 9–35.

Hedlund, G. (1993), 'Assumptions of hierarchy and heterarchy: an application to the multinational corporation', in S. Ghoshal and D.E. Westney (eds), *Organization Theory and the Multinational Corporation*, London: Macmillan.

Hedlund, G. (1999), 'The intensity and extensity of knowledge and the multinational corporation as a nearly recomposable system', *Management International Review*, **1**, 5–44.

Henry, O. (2002), 'The acquisition of symbolic capital by consultants: the French case', in M. Kipping and L. Engwall (eds), *Management Consulting: Emergence and Dynamics of a Knowledge Industry*, Oxford: Oxford University Press, pp. 19–35.

Jones, A. (2002), 'The "global city" misconceived: the myth of "global management" in transnational service firms', *Geoforum*, **33** (3), 335–50.

Jones, A. (2003), *Management Consultancy and Banking in an Era of Globalization*, London: Palgrave Macmillan.

Kipping, M. (2002), 'Trapped in their wave: the evolution of management consultancies', in T. Clark and R. Fincham (eds), *Critical Consulting: New Perspectives on the Management Advice Industry*, Oxford: Blackwell, pp. 28–49.

Kipping, M. and L. Engwall (2002), *Management Consulting: Emergence and Dynamics of a Knowledge Industry*, Oxford: Oxford University Press.

Kipping, M., S. Furusten and H. Gammelsaeter (2003), 'Converging towards American dominance? Developments and structures of consultancy fields in Europe', *Enterprises et Histoire*, **33**, 25–40.

Kristensen, P.H. and J. Zeitlin (2005), *Local Players in Global Games*, Oxford: Oxford University Press.

Lamoreaux, N.R., D.M.G. Raff and P. Temin (2002), 'Beyond markets and hierarchies: toward a new synthesis of American business history', NBER working paper 9029.

Langolis, R. and P. Robertson (1995), *Firms, Markets and Economic Change*, London and New York: Routledge.

Langlois, R.N. (2003), 'The vanishing hand: the changing dynamics of industrial capitalism', *Industrial and Corporate Change*, **12** (2), 351–85.

Lowendahl, B.R. (2000), 'The globalization of professional business service firms: fads or genuine source of competitive advantage', in Y. Aharoni and L. Nachum (eds), *Globalization of Services*, London: Routledge, pp. 142–62.

Lowendahl, B., O. Revang and S. Fosstenlokken (2001), 'Knowledge and value creation in professional service firms: a framework for analysis', *Human Relations*, **54** (7), 911–31.

McKenna, C., M.-L. Djelic and A. Ainamo (2003), 'Message and medium: the role of consulting firms in globalization and its local interpretation', in M.-L. Djelic and S. Quack (eds), *Globalization and Institutions*, Cheltenham, UK and Northampton, MA, USA: Edward Elgar, pp. 83–107.

Mills, P.K. and M. Morris (1986), 'Clients as partial employees of service organizations: role development in client participation', *Academy of Management Review*, **11** (4), 726–35.

Morgan, G. (2001a), 'Transnational communities and business systems', *Global Networks*, **1** (2), 113–30.

Morgan, G. (2001b), 'The development of transnational standards and regulations and their impacts on firms', in G. Morgan, P.H. Kristensen and R. Whitley (eds), *The Multinational Firm: Organizing Across National and Institutional Divides*, Oxford: Oxford University, pp. 225–52.

Morgan, G. (2005), 'Understanding multinational corporations', in S. Ackroyd, P. Arnold, R. Batt and P. Thompson (eds), *The Oxford Handbook of Work and Organizations*, Oxford: Oxford University Press, pp. 554–76.

Morgan, G. and S. Quack (2005a), 'Institutional legacies and firm dynamics: the growth and internationalization of UK and German law firms', *Organization Studies*, **26** (12), 1765–85.

Morgan, G. and S. Quack (2005b), 'Internationalization and capability development in professional services firms', in G. Morgan, R. Whitley and E. Moen (eds), *Changing Capitalisms? Internationalization, Institutional Change and Systems of Economic Organization*, Oxford: Oxford University Press, pp. 277–311.

Morgan, G., P.H. Kristensen and R. Whitley (eds) (2001), *The Multinational Firm: Organizing Across National and Institutional Divides*, Oxford: Oxford University Press.

Morgan, G., R. Whitley and E. Moen (eds) (2005), *Changing Capitalisms? Internationalization, Institutional Change and Systems of Economic Organization*, Oxford: Oxford University Press.

Nachum, L. (2000), 'FDI, the location advantages of countries and the competitiveness of TNCs: US FDI in professional service industries', in Y. Aharoni and L. Nachum (eds), *Globalization of Services*, London: Routledge, pp. 75–92.

Nanda, A. and K. Morrell (2004), 'Developing professionals – the BCG way (A)', Harvard Business School case 9-903-113.

Nohria, N. and N. Ghoshal (1997), *The Differentiated Network: Organizing Multinational Corporations for Value Creation*, New York: Jossey-Bass.

Pinault, L. (2001), *Consulting Demons – Inside the Unscrupulous World of Global Corporate Consulting*, New York: Harper Business.

Roberts, J. (1998), *The Internationalization of Business Service Firms*, Aldershot: Ashgate.

Roberts, J. (1999), 'The internationalization of business service firms: a stages approach', *Service Industries Journal*, **19**, 68–88.

Roberts, J. (2004), 'Global management consultancy: conceptual issues concerning the cross-border delivery of knowledge intensive services', paper presented at the British Academy of Management, September.

Rose, T. and C.R. Hinings (1999), 'Global clients' demands driving change in global business advisory firms', in D. Brock, M. Powell and C.R. Hinings (eds), *Restructuring the Professional Organization*, London: Routledge, pp. 41–67.

Rudolph, H. and J. Okech (2003), 'Computer, Köpfe, Communities of Practice. Internationales Wissensmanagement in großen Unternehmensberatungen', in C. Dörrenbächer (ed.), *Modelltransfer in Multinationalen Unternehmen*, Berlin: Edition sigma, pp. 29–52.

Rudolph, H. and J. Okech (2004), *Wer anderen einen Rat erteilt . . . Wettbewerbsstrategien und Personalpolitiken von Unternehmensberatungen in Deutschland*, Berlin: Edition sigma.

Sahlin-Andersson, K. and L. Engwall (2003), *The Expansion of Management Knowledge: Carriers, Flows and Sources*, Stanford, CA: Stanford University Press.

Sassen, S. (2001), *The Global City*, revised edn, Princeton, NJ: Princeton University Press.

Solvell, O. (2003), 'The multi-home based multinational: combining global competitiveness and local innovativeness', in J. Birkinshaw, S. Ghoshal, C. Markides, J. Stopford and G. Yip (eds), *The Future of the Multinational Company*, London: Wiley, pp. 34–44.

Solvell, O., and I. Zander (1998), 'International diffusion of knowledge: isolating mechanisms and the role of the MNE', in A. D. Chandler, O. Solvell, and P. Hagstrom (eds), *The Dynamic Firm: The Role of Technology, Strategy, Organization and Regions*, Oxford: Oxford University Press.

Spar, D. (1997), 'Lawyers abroad: the internationalization of legal practices', *California Management Review*, **39** (1), 8–28.

Stevens, M. (1981), *The Big Eight*, New York: Collier.

Sturdy, A. (1997), 'The consultancy process – an insecure business', *Journal of Management Studies*, **34** (3), 389–413.

Sturgeon, T.J. (2002), 'Modular production networks: a new American model of industrial organization', *Industrial and Corporate Change*, **11** (3), 451–96.

Taylor, P.J. (2004), *World City Network*, London: Routledge.

Teece, D. (2003), 'Expert talent and the design of (professional services) firms', *Industrial and Corporate Change*, **12** (4), 895–916.

Thrift, N. (1998), 'Virtual capitalism: the globalisation of reflexive business knowledge', in J.G. Carrier and D. Miller (eds), *Virtualism: A New Political Economy*, Oxford: Berg, pp. 161–186.

United Nations Conference on Trade and Development (UNCTAD) (2004), *World Investment Report 2004*, Geneva: UNCTAD.

Warf, B. (1997), 'Global dimensions of US legal services', *Professional Geographer*, **53** (3), 398–406.

Werr, A. (2002), 'The internal creation of consulting knowledge: a question of structuring experience', in M. Kipping and L. Engwall (eds), *Management Consulting: Emergence and Dynamics of a Knowledge Industry*, Oxford: Oxford University Press, pp. 91–108.

Werr, A. and T. Stjrenberg (2003), 'Exploring management consulting firms as knowledge systems', *Organization Studies*, **24** (6), 881–908.

Whitley, R. (2005), 'Developing transnational organizational capabilities in multinational companies: institutional constraints on authority sharing and careers in six types of MNC', in G. Morgan, R. Whitley and E. Moen (eds), *Changing Capitalisms? Internationalization, Institutional Change and Systems of Economic Organization*, Oxford: Oxford University Press, pp. 235–76.

Index